Process–Structure–Properties in Polymer Additive Manufacturing II

Process–Structure–Properties in Polymer Additive Manufacturing II

Editors

Swee Leong Sing
Wai Yee Yeong

MDPI • Basel • Beijing • Wuhan • Barcelona • Belgrade • Manchester • Tokyo • Cluj • Tianjin

Editors
Swee Leong Sing
Department of Mechanical Engineering
National University of Singapore
Singapore
Singapore

Wai Yee Yeong
Singapore Centre for 3D Printing, School of Mechanical and Aerospace Engineering
Nanyang Technological University
Singapore
Singapore

Editorial Office
MDPI
St. Alban-Anlage 66
4052 Basel, Switzerland

This is a reprint of articles from the Special Issue published online in the open access journal *Polymers* (ISSN 2073-4360) (available at: www.mdpi.com/journal/polymers/special_issues/process_structure_polym_additive_manufacturing_II).

For citation purposes, cite each article independently as indicated on the article page online and as indicated below:

LastName, A.A.; LastName, B.B.; LastName, C.C. Article Title. *Journal Name* **Year**, *Volume Number*, Page Range.

ISBN 978-3-0365-4484-7 (Hbk)
ISBN 978-3-0365-4483-0 (PDF)

© 2022 by the authors. Articles in this book are Open Access and distributed under the Creative Commons Attribution (CC BY) license, which allows users to download, copy and build upon published articles, as long as the author and publisher are properly credited, which ensures maximum dissemination and a wider impact of our publications.

The book as a whole is distributed by MDPI under the terms and conditions of the Creative Commons license CC BY-NC-ND.

Contents

About the Editors vii

Swee Leong Sing and Wai Yee Yeong
Recent Progress in Research of Additive Manufacturing for Polymers
Reprinted from: *Polymers* **2022**, *14*, 2267, doi:10.3390/polym14112267 1

Orhan Gülcan, Kadir Günaydın and Aykut Tamer
The State of the Art of Material Jetting—A Critical Review
Reprinted from: *Polymers* **2021**, *13*, 2829, doi:10.3390/polym13162829 3

Balakrishnan Nagarajan, Yingnan Wang, Maryam Taheri, Simon Trudel, Steven Bryant and Ahmed Jawad Qureshi et al.
Development and Characterization of Field Structured Magnetic Composites
Reprinted from: *Polymers* **2021**, *13*, 2843, doi:10.3390/polym13172843 23

Changlang Wu, Truong Tho Do and Phuong Tran
Mechanical Properties of PolyJet 3D-Printed Composites Inspired by Space-Filling Peano Curves
Reprinted from: *Polymers* **2021**, *13*, 3516, doi:10.3390/polym13203516 39

Razvan Udroiu
New Methodology for Evaluating Surface Quality of Experimental Aerodynamic Models Manufactured by Polymer Jetting Additive Manufacturing
Reprinted from: *Polymers* **2022**, *14*, 371, doi:10.3390/polym14030371 61

Asmak Abdul Samat, Zuratul Ain Abdul Hamid, Mariatti Jaafar and Badrul Hisham Yahaya
Mechanical Properties and In Vitro Evaluation of Thermoplastic Polyurethane and Polylactic Acid Blend for Fabrication of 3D Filaments for Tracheal Tissue Engineering
Reprinted from: *Polymers* **2021**, *13*, 3087, doi:10.3390/polym13183087 81

Zihui Zhang, Fengtai He, Bo Wang, Yiping Zhao, Zhiyong Wei and Hao Zhang et al.
Biodegradable PGA/PBAT Blends for 3D Printing: Material Performance and Periodic Minimal Surface Structures
Reprinted from: *Polymers* **2021**, *13*, 3757, doi:10.3390/polym13213757 97

Dorin-Ioan Catana, Mihai-Alin Pop and Denisa-Iulia Brus
Comparison between Tests and Simulations Regarding Bending Resistance of 3D Printed PLA Structures
Reprinted from: *Polymers* **2021**, *13*, 4371, doi:10.3390/polym13244371 111

Fengze Jiang and Dietmar Drummer
Analysis of UV Curing Strategy on Reaction Heat Control and Part Accuracy for Additive Manufacturing
Reprinted from: *Polymers* **2022**, *14*, 759, doi:10.3390/polym14040759 123

Yanis Abdelhamid Gueche, Noelia M. Sanchez-Ballester, Bernard Bataille, Adrien Aubert, Jean-Christophe Rossi and Ian Soulairol
Investigating the Potential Plasticizing Effect of Di-Carboxylic Acids for the Manufacturing of Solid Oral Forms with Copovidone and Ibuprofen by Selective Laser Sintering
Reprinted from: *Polymers* **2021**, *13*, 3282, doi:10.3390/polym13193282 135

Samuel Schlicht, Sandra Greiner and Dietmar Drummer
Low Temperature Powder Bed Fusion of Polymers by Means of Fractal Quasi-Simultaneous
Exposure Strategies
Reprinted from: *Polymers* **2022**, *14*, 1428, doi:10.3390/polym14071428 . . . . **155**

About the Editors

Swee Leong Sing

Swee Leong Sing is an Assistant Professor in the Department of Mechanical Engineering, National University of Singapore (NUS). Prior to joining NUS in August 2021, he was a Presidential Postdoctoral Fellow at the Singapore Centre for 3D Printing (SC3DP) and School of Mechanical and Aerospace Engineering (MAE), Nanyang Technological University (NTU), Singapore after being awarded the prestigious fellowship in 2020.

Swee Leong has been active in the 3D printing field for more than 8 years. He obtained his BEng (Hons) in Aerospace Engineering and PhD in Mechanical Engineering with a topic in additive manufacturing (AM) in 2012 and 2016, respectively. Swee Leong's research focuses on using advanced manufacturing techniques as enablers for materials development and to create strategic values for the industries. He is also active in inter-disciplinary research and translational work. His research has been awarded the Best PhD Thesis Award by MAE, NTU, Singapore, as well as the Springer Theses Award from Springer Nature, Germany in 2017. Swee Leong has worked on numerous 3D printing projects with government agencies, universities, research institutes and industrial collaborators. Swee Leong has filed five patents pertaining to 3D printing processes and materials. He has published 1 book, 4 book chapters and more than 50 peer reviewed journal and conference articles. Swee Leong currently has an h-index of 30 with more than 4500 citations (Web of Science, 31 May 2022).

Wai Yee Yeong

Associate Professor Wai Yee Yeong is the winner of the TCT Women in 3D Printing Innovator Award 2019 and named as one of the Singapore 100 Women in Tech List 2021. Her work is well-recognized, with an H-index of 56 and more than 12,000 citations on Google Scholar. She has filed multiple patents and knowhows, with a key interest in Bioprinting and the 3D printing of new materials. On academic fronts, she is the Associate Editor for 2 international journals and has authored 3 textbooks on 3D printing. Prof Yeong serves as Program Director in the Singapore Centre for 3D Printing and HP-NTU Digital Manufacturing Corp Lab. Prof Yeong was awarded the NRF Investigatorship (Class of 2022) in her pursuit for groundbreaking and high-risk research.

Editorial

Recent Progress in Research of Additive Manufacturing for Polymers

Swee Leong Sing [1,*] and Wai Yee Yeong [2]

1. Department of Mechanical Engineering, National University of Singapore, Singapore 117575, Singapore
2. Singapore Centre for 3D Printing, School of Mechanical & Aerospace Engineering, Nanyang Technological University, Singapore 639798, Singapore; wyyeong@ntu.edu.sg
* Correspondence: sing0011@e.ntu.edu.sg or sweeleong.sing@nus.edu.sg

Citation: Sing, S.L.; Yeong, W.Y. Recent Progress in Research of Additive Manufacturing for Polymers. *Polymers* 2022, 14, 2267. https://doi.org/10.3390/polym14112267

Received: 12 May 2022
Accepted: 30 May 2022
Published: 2 June 2022

Publisher's Note: MDPI stays neutral with regard to jurisdictional claims in published maps and institutional affiliations.

Copyright: © 2022 by the authors. Licensee MDPI, Basel, Switzerland. This article is an open access article distributed under the terms and conditions of the Creative Commons Attribution (CC BY) license (https://creativecommons.org/licenses/by/4.0/).

Additive manufacturing (AM) methods have grown and evolved rapidly in recent years. AM for polymers is particularly exciting and has great potential in transformative and translational research in many fields, such as biomedical [1–3], aerospace [4,5], and electronics [6,7]. Current methods for polymer AM include material extrusion, material jetting, vat photopolymerization, and powder bed fusion. As these techniques matured and developed, more functionalities have been added to AM parts. Such functionalities include multi-material fabrication [8–10] and integration with artificial intelligence [11]. These have resulted in polymer AM to evolve from a rapid prototyping tool to actual manufacturing solution.

In this special issue, state-of-the art research and review articles are collected. They focus on the process–structure–properties relationships in polymer AM. In total, one review and nine original research articles are included. Gülcan et al. provided a comprehensive review on the material jetting technique for polymer AM by analyzing the effect of the critical process parameters and providing benchmarking with other manufacturing processes [12]. In their research, Nagarajan et al. investigated the use of polymer composites that contain ferromagnetic fillers for applications in electronic and electrical devices. These composites were processed using material jetting and alignment of the fillers was achieved using magnetic field [13]. Wu et al. also used material jetting to produce novel composite materials that are multi-material [14]. Udroiu studied the use of material jetting produced surfaces for aerodynamic models [15]. Samat et al. evaluated the mechanical and in vitro properties of material extruded thermoplastic polyurethane and polylactic acid blend for tracheal tissue engineering [16]. Zhang et al. also used material extrusion of blends for their experiments. They studied biodegradable polyesters and adjusted the blend compositions to tailor the mechanical performance [17]. Catana et al. studied the bending resistance of polylactic acids and compared them to the simulations. They found that the AM parts deviated from simulations due to fluctuations in process parameters [18]. Jiang and Drummer studied the effect of curing strategy on the part accuracy produced by vat photopolymerization [19]. Gueche et al. investigated the feasibility of using dicarboxylic acids to produce solid oral forms with copovidone and ibuprofen using powder bed fusion [20]. Finally, Schlicht et al. developed new scanning strategies using quasi-simultaneous exposure of fractal scan paths for powder bed fusion of polymers that can reduce the energy consumption of the process [21].

Acknowledgments: This research is supported by the National Research Foundation, Prime Minister's Office, Singapore under its Medium-Sized Centre funding scheme.

Conflicts of Interest: The authors declare no conflict of interest.

References

1. Luis, E.; Pan, H.M.; Bastola, A.K.; Bajpai, R.; Sing, S.L.; Song, J.; Yeong, W.Y. 3D Printed Silicone Meniscus Implants: Influence of the 3D Printing Process on Properties of Silicone Implants. *Polymers* **2020**, *12*, 2136. [CrossRef]
2. Luis, E.; Pan, H.M.; Sing, S.L.; Bajpai, R.; Song, J.; Yeong, W.Y. 3D Direct Printing of Silicone Meniscus Implant Using a Novel Heat-Cured Extrusion-Based Printer. *Polymers* **2020**, *12*, 1031. [CrossRef] [PubMed]
3. Khan, Z.N.; Albalawi, H.I.; Valle-Pérez, A.U.; Aldoukhi, A.; Hammad, N.; de León, E.H.-P.; Abdelrahman, S.; Hauser, C.A.E. From 3D printed molds to bioprinted scaffolds: A hybrid material extrusion and vat polymerization bioprinting approach for soft matter constructs. *Mater. Sci. Addit. Manuf.* **2022**, *1*, 7.
4. Wang, F.; Zheng, J.; Wang, G.; Jiang, D.; Ning, F. A novel printing strategy in additive manufacturing of continuous carbon fiber reinforced plastic composites. *Manuf. Lett.* **2021**, *27*, 72–77. [CrossRef]
5. Weyhrich, C.W.; Long, T.E. Additive manufacturing of high-performance engineering polymers: Present and future. *Polym. Int.* **2021**, *71*, 532–536. [CrossRef]
6. Criado-Gonzalez, M.; Dominguez-Alfaro, A.; Lopez-Larrea, N.; Alegret, N.; Mecerreyes, D. Additive Manufacturing of Conducting Polymers: Recent Advances, Challenges, and Opportunities. *ACS Appl. Polym. Mater.* **2021**, *3*, 2865–2883. [CrossRef]
7. Divakaran, N.; Das, J.P.; V, A.K.P.; Mohanty, S.; Ramadoss, A.; Nayak, S.K. Comprehensive review on various additive manufacturing techniques and its implementation in electronic devices. *J. Manuf. Syst.* **2022**, *62*, 477–502. [CrossRef]
8. Ng, W.L.; Ayi, T.C.; Liu, Y.-C.; Sing, S.L.; Yeong, W.Y.; Tan, B.-H. Fabrication and Characterization of 3D Bioprinted Triple-layered Human Alveolar Lung Models. *Int. J. Bioprint.* **2021**, *7*, 332. [CrossRef]
9. Lee, J.M.; Sing, S.L.; Yeong, W.Y. Bioprinting of Multimaterials with Computer-aided Design/Computer-aided Manufacturing. *Int. J. Bioprint.* **2020**, *6*, 245. [CrossRef]
10. Jiang, H.; Aihemaiti, P.; Aiyiti, W.; Kasimu, A. Study Of the compression behaviours of 3D-printed PEEK/CFR-PEEK sandwich composite structures. *Virtual Phys. Prototyp.* **2022**, *17*, 138–155. [CrossRef]
11. Goh, G.D.; Sing, S.L.; Lim, Y.F.; Thong, J.L.J.; Peh, Z.K.; Mogali, S.R.; Yeong, W.Y. Machine learning for 3D printed multi-materials tissue-mimicking anatomical models. *Mater. Des.* **2021**, *211*, 110125. [CrossRef]
12. Gülcan, O.; Günaydın, K.; Tamer, A. The State of the Art of Material Jetting—A Critical Review. *Polymers* **2021**, *13*, 2829. [CrossRef] [PubMed]
13. Nagarajan, B.; Wang, Y.; Taheri, M.; Trudel, S.; Bryant, S.; Qureshi, A.J.; Mertiny, P. Development and Characterization of Field Structured Magnetic Composites. *Polymers* **2021**, *13*, 2843. [CrossRef] [PubMed]
14. Wu, C.; Do, T.T.; Tran, P. Mechanical Properties of PolyJet 3D-Printed Composites Inspired by Space-Filling Peano Curves. *Polymers* **2021**, *13*, 3516. [CrossRef] [PubMed]
15. Udroiu, R. New Methodology for Evaluating Surface Quality of Experimental Aerodynamic Models Manufactured by Polymer Jetting Additive Manufacturing. *Polymers* **2022**, *14*, 371. [CrossRef]
16. Samat, A.A.; Hamid, Z.A.A.; Jaafar, M.; Yahaya, B.H. Mechanical Properties and In Vitro Evaluation of Thermoplastic Polyurethane and Polylactic Acid Blend for Fabrication of 3D Filaments for Tracheal Tissue Engineering. *Polymers* **2021**, *13*, 3087. [CrossRef]
17. Zhang, Z.; He, F.; Wang, B.; Zhao, Y.; Wei, Z.; Zhang, H.; Sang, L. Biodegradable PGA/PBAT Blends for 3D Printing: Material Performance and Periodic Minimal Surface Structures. *Polymers* **2021**, *13*, 3757. [CrossRef]
18. Catana, D.-I.; Pop, M.-A.; Brus, D.-I. Comparison between Tests and Simulations Regarding Bending Resistance of 3D Printed PLA Structures. *Polymers* **2021**, *13*, 4371. [CrossRef]
19. Jiang, F.; Drummer, D. Analysis of UV Curing Strategy on Reaction Heat Control and Part Accuracy for Additive Manufacturing. *Polymers* **2022**, *14*, 759. [CrossRef]
20. Gueche, Y.A.; Sanchez-Ballester, N.M.; Bataille, B.; Aubert, A.; Rossi, J.-C.; Soulairol, I. Investigating the Potential Plasticizing Effect of Di-Carboxylic Acids for the Manufacturing of Solid Oral Forms with Copovidone and Ibuprofen by Selective Laser Sintering. *Polymers* **2021**, *13*, 3282. [CrossRef]
21. Schlicht, S.; Greiner, S.; Drummer, D. Low Temperature Powder Bed Fusion of Polymers by Means of Fractal Quasi-Simultaneous Exposure Strategies. *Polymers* **2022**, *14*, 1428. [CrossRef] [PubMed]

Review

The State of the Art of Material Jetting—A Critical Review

Orhan Gülcan [1,*], Kadir Günaydın [1,*] and Aykut Tamer [2]

1 General Electric Aviation, Gebze 41400, Kocaeli, Turkey
2 Department of Mechanical Engineering, Imperial College London, London SW7 2AZ, UK; a.tamer@imperial.ac.uk
* Correspondence: orhan.gulcan@ge.com (O.G.); kadir.gunaydin@ge.com (K.G.); Tel.: +90-262-677-8410

Abstract: Material jetting (MJ) technology is an additive manufacturing method that selectively cures liquid photopolymer to build functional parts. The use of MJ technology has increased in popularity and been adapted by different industries, ranging from biomedicine and dentistry to manufacturing and aviation, thanks to its advantages in printing parts with high dimensional accuracy and low surface roughness. To better understand the MJ technology, it is essential to address the capabilities, applications and the usage areas of MJ. Additionally, the comparison of MJ with alternative methods and its limitations need to be explained. Moreover, the parameters influencing the dimensional accuracy and mechanical properties of MJ printed parts should be stated. This paper aims to review these critical aspects of MJ manufacturing altogether to provide an overall insight into the state of the art of MJ.

Keywords: tray location; build direction; surface finish; matte; glossy

1. Introduction

The demand for complex parts is steadily increasing in different industries (especially in aerospace, automotive and biomedical industries) to manufacture lighter parts with higher stiffness, higher strength and lower cost. Thanks to the recent advances in additive manufacturing (AM) technologies, engineers have more freedom to design and produce complex parts which were more difficult if not impossible to manufacture with conventional means [1]. The main difference in AM from conventional, subtractive manufacturing methods is that it is based on a layer-by-layer manufacturing which results in a reduced low buy to fly ratio (the ratio of weight of raw material to weight of the final part) [2,3].

According to "The American Society for Testing and Materials (ASTM) Committee F42 on Additive Manufacturing Technologies" AM technologies can be classified as: powder bed fusion, material jetting, vat photopolymerization, directed energy deposition, material extrusion, binder jetting and sheet lamination [4]. The two different terms are utilized to refer to material jetting processes synonymously. These names are used due to secured naming rights of the material jetting printer manufacturers Stratasys (PolyJet) and 3DSystems (MultiJet). The technology was first developed by Objet Geometries Ltd. in 2000 and was acquired later by Stratasys in 2012 [5]. According to ISO/ASTM 52900: 2015 standard, "droplets of feedstock material are selectively deposited" in MJ technology [6]. Although the MJ printer design varies slightly from manufacturer to manufacturer, a general schematic representation of MJ can be seen in Figure 1. In MJ, air-excluding tanks are used to store photopolymer materials and these are deposited as droplets forming a very thin layer on the build platform after heating photopolymer in the transmission line in which photopolymer is transmitted from tank to nozzle [7]. Ultraviolet (UV) light is emitted onto the molten material on the build platform for curing. In this photopolymerization/photo-curing process, a light source of a specific wavelength is used to cure monomers/oligomers in the liquid state [8]. Unlike the wavelength of lamps used by SLA (355 nm) and DLP (405 nm), the wavelength of the light source in MJ can

Citation: Gülcan, O.; Günaydın, K.; Tamer, A. The State of the Art of Material Jetting—A Critical Review. *Polymers* **2021**, *13*, 2829. https://doi.org/10.3390/polym13162829

Academic Editor: Swee Leong Sing

Received: 17 July 2021
Accepted: 13 August 2021
Published: 23 August 2021

Publisher's Note: MDPI stays neutral with regard to jurisdictional claims in published maps and institutional affiliations.

Copyright: © 2021 by the authors. Licensee MDPI, Basel, Switzerland. This article is an open access article distributed under the terms and conditions of the Creative Commons Attribution (CC BY) license (https://creativecommons.org/licenses/by/4.0/).

theoretically be unrestricted [9], but, practically, a light source of a wavelength between 190 and 400 nm is used [10,11]. After curing a layer, the build platform is lowered at a level of certain layer thickness amount and new liquid material is jetted onto the previous layer. After curing each successive layer, a full-scale part is completely built [12]. Since liquid or molten material is used in MJ, a gel-like support structure is needed, especially in overhang regions. These support structures are removed from the part using different methods: sonication in a bath of sodium hydroxide solution, heating or using a high-pressure water jet [13].

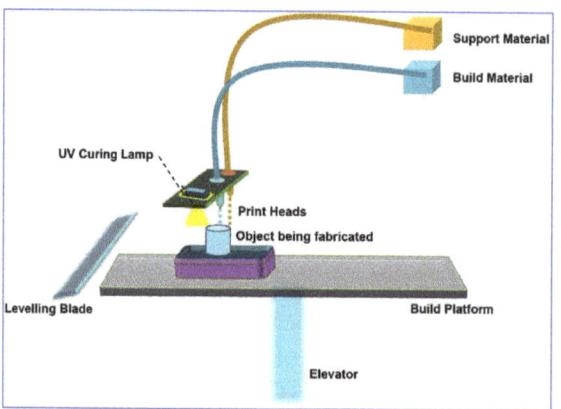

Figure 1. A schematic representation of MJ [14].

The MJ technology is a prominent additive manufacturing method in the polymer printing field due its advantages in comparison with other polymer printing techniques. Technology allows us to adapt thin layer thicknesses which allows for printing high-quality parts and less evident staircase effects and thin wall features [8]. The layer thicknesses can be as low as 16 μm [14]. The low surface roughness texture is another advantage which is one of the major problems for additive manufacturing technologies. There are two surface finish options in MJ technology: matte and glossy. In matte the setting, the whole part is covered with support material. In the glossy setting, only structurally needed areas are supported and the model is exposed to air during curing. After printing, supported areas will be matte and unsupported areas will be glossy. Furthermore, no post-processes are required for MJ technology, and parts are usable in the as-built condition after separating them from build platform and support removal processes. Rather than printing parts directly on substrate, a photopolymer resin can be printed onto substrate as a printing bed [15]. Due to easy detachment of photopolymer, separating printed parts from the built plate is possible with a hand tool by applying less amount of force. In addition, different materials such as PLA, ABS, polyamide and their combinations can be combined in a single in MJ technology, called the multi-material approach [14]. The multi-material approach can be utilized to produce composite parts for specific purposes. Lastly, MJ printers have a closed ambient chamber for production which prevents undesirable effects of draught or dirt and can be used in offices and homes.

In this study, the state of the art of MJ technology was reviewed. To better understand the MJ technology, its capabilities, applications and usage areas were addressed. Additionally, MJ was compared with other technologies and its limitations were examined. Moreover, the parameters influencing the dimensional accuracy and mechanical properties of MJ printed parts were investigated. To the authors' knowledge, no other review has addressed these aspects of MJ in the literature. Section 2 presents process and materials where polymeric composites are investigated and the effect of tray location, build orientation and surface finish setting on dimensional accuracy, surface roughness and mechanical properties of MJ printed parts are reviewed. In this section, MJ capability and performance of MJ printed parts and comparison of MJ technology with other technologies are also investigated. Section 3 reports the application of MJ technology in different industries. The paper is concluded by a summary of the findings.

2. Process and Materials

2.1. Materials

In MJ, thermoplastic, thermosetting and elastomeric polymers with different mechanical properties are used. Pilipović et al. compared mechanical properties of FullCure, VeroBlue and VeroBlack materials in MJ. They stated that FullCure material has the maximum flexural and tensile strength, followed by VeroBlue and VeroBlack. They also mentioned that FullCure material resulted in the lowest surface roughness [16]. O'Neill et al. compared mechanical properties of different types of materials in MJ. They stated that elastic modulus, strain at fracture, maximum compressive strength and yield strength of MED610 is higher than Vero White and RGD525. It was found that RGD 525 has the lowest and MED610 has the highest wear rate [13]. Kaweesa and Meisel designed fatigue test bars in which the middle sections have a material gradient transition region with continuous gradients and stepwise gradient types of functionally graded material interface designs with two materials: VeroCyan and TangoBlackPlus. They found that stepwise gradient specimens with a longer flexible material region showed interfacial failure and higher fatigue life than continuous gradient specimens and fatigue life increased with a decrease in material gradient transition lengths [17].

2.2. Polymeric Composites via MJ

The improvement in AM enables new ideas to produce innovative parts to fulfill mechanical needs. One of these ideas is the production of composite materials, which was inspired by natural composites, with a mixture of soft–hard materials to create new materials with better mechanical properties [18]. The combination of two different materials exhibiting different hardness specifications creates unique properties such as improved flexibility and hardness [19,20]. The benefit of AM methods for the production of these complex structures is relatively less time needed and cost in comparison to conventional techniques. These composite materials are generally utilized in applications of specially designed structures such as hierarchical structures, including honeycombs, lattice structures and foams for the purpose of energy absorption [21–23]. MJ is one of the most convenient methods for composite material production. Sugawaneswaran et al. suggested a novel methodology for fabrication of a randomly oriented plastic reinforced composite structure with elastomer matrix phase. The random distribution of reinforcing materials was determined in a CAD model. As a result, the composite exhibited 22% more stiffness and 10% more elongation in comparison with constituent materials, while the orientation of reinforcements did not inflences the stress–strain curves considerably [24,25]. One of the composite structures that can be fabricated with MJ is interpenetrating phase composites (IPCs). An IPC is made of lattice-based solid sheets embedded in a soft matrix. Dalag et al. investigated the compression behavior of triply periodic minimal surface (TPMS) reinforced IPCs. It was concluded that an increase in reinforcement structure volume fraction of 5% to 20% affects the mechanical behavior of the IPC. Moreover, with the increase in the reinforcement structure volume fraction, ultimate compression strength and yield strength rise considerably. During the deformation of the IPCs, local bucklings and debonding failure mechanisms were experienced [26]. As for the dimensional accuracy of 3D printed microcomposites via MJ, Tee et al. [27] suggested that geometric resolution is convenient whenever it is greater than 500 µm. Additionally, in this study, the mechanical behaviors of 3D printed polymeric microcomposites with different compositions and arrangements of reinforced particles are investigated. As a result, it is shown that orientations and reinforced particle geometries dominate the stiffness of the composites under compressive loads. However, the composite specimen tensile test results showed that build orientation is insignificant for strength whenever the reinforced particle volume fraction is kept at a level of 5%. In addition, it was stated that the composition and reinforced particle arrangement considerably influences the mechanical behavior. In particular, the hardness of the reinforced particles has an effect on the failure mechanism of the polymeric composite structures. Even though the use of hard particles is able to increase the strength, soft particles serve as crack initiators.

2.3. The Effect of Parameters

Dimensional accuracy, surface roughness and mechanical properties of MJ printed parts are crucial to obtain correct and reliable results after measurements and proper functioning in the final assembly. Tray location, layer thickness, build orientation, surface finish, material type and post-processing are the most important parameters that affect dimensional accuracy, surface roughness and mechanical properties of MJ manufactured parts [28,29].

In the literature, the percentage or level of impact of these parameters on dimensional accuracy, surface roughness and mechanical properties has been investigated. Related to surface roughness, Kechagias et al. investigated the effect of layer thickness, surface finish setting and scale factor on the surface roughness of MJ printed parts. They stated that a smaller layer thickness (16 µm) and glossy surface finish setting gives the best results in terms of surface roughness. They also indicated that scale factor was not a dominant factor for surface roughness [30]. Aslani et al. investigated the effect of layer thickness, surface finish setting (matte or glossy) and build scale on surface roughness of MJ printed parts. They observed that the surface finish setting has the greatest effect on surface roughness with a contribution rate of 95%. On the contrary, the layer thickness and the build scale have very little effect on surface roughness with a contribution rate of less than 3% [31].

In MJ, dimensional accuracy of printed parts depends mostly on polymerization speed and photocurable resin material viscosity [32]. Kechagias et al. investigated the effect of layer thickness, surface finish setting (matte or glossy) and model scale factor on dimensional accuracy of internal and external features produced by MJ. Their results showed that layer thickness and model scale factor were the two main factors affecting dimensional accuracy of internal features. For external features, all factors were equally important, although layer thickness has higher importance in the X and Y direction and the model scale factor has higher importance in the Z direction [33]. Aslani et al. investigated the effect of layer thickness, surface finish setting and scale on dimensional accuracy of MJ 3D printed parts using the grey Taguchi method. They stated that scale had the largest impact with contribution of 66.54%, followed by build style (contribution = 33.16%) and layer thickness (contribution = 0.3%) [34].

Pugalendhi et al. investigated the effect of finish option, material, thickness and shape on build time, specimen height and number of layers in MJ printed parts. They used Taguchi analysis to perform DOE and ANOVA to determine the % contribution of each factor to the results. They suggested that for all measured values, thickness was the most dominant factor [35].

2.3.1. Tray Location

In MJ, tray location can be defined in terms of the X, Y and Z axis. The jetting head moves along the X axis and along the transverse Y axis along which jetting orifices are located in parallel. After each layer, the build plate moves along the Z axis (Figure 2) [36]. Locating the part along the X or Y axis of the build tray in MJ has a considerable effect on the mechanical properties and surface roughness of final parts. As stated by Barclift et al., the distribution of the same parts on the build platform can affect mechanical properties due to over curing of some parts in different locations while other parts are being cured. They also stated that decreasing the part spacing increased the part strength [37]. Pilipović et al. stated that dimensional accuracy and repeatability on the X and Y axis is better than it is for the Z axis based on distance measurements in MJ technology [8]. Cazon et al. stated that the best roughness results were obtained when standard tensile specimens (BS EN

ISO 527-2:1996) were placed close to the XY plane (in which the first letter designates the specimen's main axis and the second letter is the minor axis), in other words, parts printed along the X axis gave the best results in terms of stiffness [29]. Gay et al. stated that part spacing on the X axis and surface roughness of parts had no significant influence on mechanical properties but part spacing on the Y axis has a considerable effect on properties. Therefore, the authors suggested placing the parts on the Y axis as close as possible to obtain better mechanical properties. They also stated that orientation of the part affected the mechanical properties slightly and maximum properties were observed when the parts were located parallel to the coordinate axis [36]. Beltrán et al. investigated the effect of orientation, location and size of the part within the build platform on dimensional and geometrical accuracy of cylindrical features. They stated that part orientation and part size have the most influence and part location within the build platform has a relatively lower influence on accuracy of these parts [38].

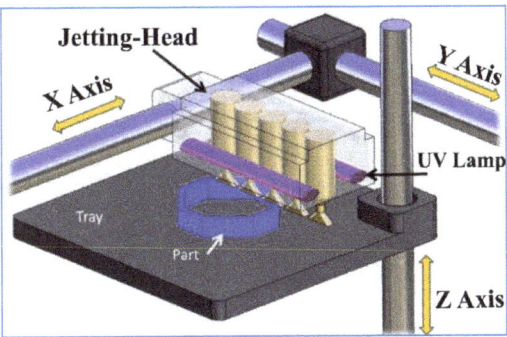

Figure 2. X, Y and Z axis in MJ [36].

2.3.2. Build Orientation

The build orientation of parts with respect to the build tray affects surface roughness, dimensional accuracy and mechanical properties. To obtain better values of these characteristics, it is important to know their effect on the properties of the final part. Kumar and Kumar stated that the surface roughness generally increased as build orientation increased until 90° (build orientation with respect to the build direction z) in MJ printed parts [39]. Kumar and Kumar investigated the effect of finish type, local surface orientation and layer thickness on surface roughness of FullCure 720 and VeroBlue 840 printed parts by MJ. They stated that the major factors affecting surface roughness were surface orientation and finish type, and maximum surface roughness was obtained for the 90° surface angle [40]. Kechagias and Stavropoulos reported that when the angle of sloped surfaces of specimens was increased, surface roughness values were also increased. The best and worst values of surface roughness were achieved when the angle of sloped surfaces was zero and 90°, respectively [41]. Khoshkhoo et al. investigated the effect of build orientation and printing direction on asymmetry, rough surface, stair-stepping effects and traces of flowed material along the surface. They suggested that for achieving better surface finish, design fidelity and dimensional accuracy, the build platform can be tilted and printing parameters (build orientation, support strategy) can be changed [42]. Vidakis et al. investigated the effect of sloped surface angle of the part on surface roughness of MJ printed parts. They used 0, 15, 30, 45, 60, 75 and 90° slope values and the X and Y direction as experimental inputs. Results showed that parts produced in the X direction showed better surface roughness when the sloped surface angle was below 45° and, after that value, parts produced in the Y direction had lower surface roughness values [43].

Kent et al. investigated the effect of build orientation on surface roughness and dimensional accuracy of MJ printed parts with four different materials. They found that build orientation has a very high impact on these measurements, but the results were not fully consistent with each other [32]. Haghighi et al. stated that a horizontal build resulted in better dimensional accuracy than vertical builds [44]. Kitsakis et al. investigated the tolerance values in the X, Y and Z direction of MJ printed parts. They reported that accuracy on the Z axis is smaller than on the X and Y axis [45,46].

Kesy and Kotlinski investigated the effect of build orientation on mechanical properties of MJ manufactured parts. In different parts built in different orientations, they found considerable changes in mechanical properties and attributed these changes to variations in the amount of UV energy that reaches the different zones on the build platform for each part [47]. Das et al. studied the effect of build orientation on tensile strength of MJ printed parts. They built the specimens in flat and inclined conditions and X, Y and 45° angle directions. They suggested that specimens built in the X direction and flat conditions showed the highest tensile properties and specimens built in the Y direction and inclined condition showed the lowest tensile properties [48]. Tomar et al. stated that building direction plays an important role in tensile properties of MJ printed parts. Their experimental study revealed that samples produced parallel to the XY plane showed higher values of yield strength, breaking stress and strain at fracture than samples produced parallel to the Z direction [49].

The glass transition temperature is one of the most important properties of polymeric materials and this property is relatively unknown for MJ printed parts. Sanders et al. investigated the effect of build orientation and layer thickness on glass transition temperature in MJ and stated that parts printed in the X direction and with larger layer thickness values demonstrated higher glass transition temperature values [50].

Blanco et al. investigated the effect of build direction on the relaxation modulus of MJ printed parts. They concluded that parts built at a 0° slope angle had the highest relaxation modulus; it decreased progressively with a 60°–75° slope angle, then it increased up to a 90° slope angle. According to the authors, a shielding effect from UV curing by the support material could be the main cause of the phenomenon [51]. Reichl et al. looked into the viscoelastic properties of MJ printed parts characterized by a complex modulus which depends on frequency and temperature. They stated that build orientation (vertical or horizontal) had no effect on the complex modulus results [52].

2.3.3. Surface Roughness Options

As mentioned in Introduction, there are two surface finish options in MJ: matte and glossy. The difference between these two settings is that, in the matte setting, the whole part is covered with support material. On the other hand, only structurally needed areas are supported and the model is exposed to air during the curing phase in the glossy setting. Yap et al. stated that when the parts were printed along the Z axis and in the matte setting, higher dimensional accuracy was obtained. On the other hand, the glossy setting reduces the need for support and cost [53]. In MJ technology, there are also two printing modes: high speed and high quality. Pugalendhi et al. investigated the mechanical properties of MJ printed parts from VeroBlue material. They stated that the high-speed printing mode is superior in terms of tensile strength, elongation at break, flexural strength and shore hardness when compared to high-quality printing mode. They also suggested that a glossy finish had lower peaks and valleys and resulted in better surface finish than matte finish options [54]. Pugalendhi et al. compared mechanical properties of VeroWhitePlus and VeroClear and stated that for both materials, glossy finish specimens give better results when compared to matte finish specimens [55]. Cazon et al. stated that the glossy setting gives better surface roughness than the matte setting [29]. Kampker et al. expressed that surface finish has a significant effect on the tensile strength and glossy surface finish parts have higher tensile strength values [56]. Moore et al. presented that glossy finish parts have higher fatigue life when compared to matte finish parts in MJ due to the lower surface roughness obtained with the glossy setting [57].

2.4. MJ Capability and Performance of MJ Printed Parts

For an effective design, dimensional accuracy and process capability of related technology are paramount. MJ technology is capable of manufacturing dimensionally accurate parts. It can be used in mass production in a short period of time with tolerance values matching IT10 grade for linear dimensions on the Z axis and ISO 286 IT9 grade for radial dimensions [58]. MJ is capable of manufacturing thin walls with a tolerance value of 25–50 μm [12]. By proper selection of parameters, dimensional accuracy tolerances can be lowered to 15 μm [38]. Silva et al. stated that microfeatures larger than 423 μm can be successfully built with MJ and large distortions or printing failures were observed below this value [59]. However, it has some limitations. Holes with diameters smaller than 0.5 mm cannot be horizontally or vertically printed and those with diameters of nearly 1 mm may have some circularity deviations in MJ technology [60]. Tee et al. stated that when parts are printed in multiple materials, MJ is capable of printing microcomposites as small as 62.5 μm but these small parts have high dimensional variations (20% to 75% dimensional variation was observed for parts with length, height and diameter features of between 62.5 μm and 250 μm) [27].

For optimal printing, the liquid material must have enough viscosity, which is generally achieved by heating the material up to 30-70 °C [61]. Yap et al. stated that thin wall features need to be oriented 0° along the Y direction (see Figure 2 for the convention of the directions) and greater than 0.4 mm for a successful build and best accuracy in width and height [53].

Wear performance of MJ printed parts has also been studied. Dangnan et al. investigated the wear and friction mechanism of MJ printed ABS and Verogray polymers. They applied different contact loads (1, 5 and 10N) and stated that in MJ parts, the wear rate depends on the applied load and surface orientation to the sliding direction. For both parallel and perpendicular orientations, the highest coefficient of friction was observed under a 1N load [62]. It was stated that with plastic reinforcements, the elastic modulus of elastomeric parts produced by MJ can be increased by 6.79%–21.03% [24,25].

In the literature, different layer thicknesses and main and support materials have been used in different MJ machines. Some of these studies are summarized in Table 1.

Table 1. Materials, machines and layer thicknesses used in the literature.

Main Material	Support Material	Machine	Layer Thickness	Reference
VeroBlack	FullCure 705	Objet Connex 350	-	[8]
VeroWhite MED610 RGD 525	FullCure 705	Objet 260 Connex1	-	[13]
VeroMagenta	-	Objet 500 Connex3	0.03 mm	[63]
VeroClear	-	-	15 μm	[64]
MED610	-	Objet Eden 260VS	16.5 μm	[65]
FullCure 720, VeroWhite, VeroBlue	-	Objet Eden 260	-	[66]
VeroWhite	-	Objet 500	16 μm	[67]
VeroWhite, FullCure 720, ABS-like	-	Objet J750	14–27 μm	[68]
VeroClear	SUP707	Objet Eden 260VS	-	[69]
VeroBlue	FullCure 705	Objet Eden 350	16 μm	[58]
VeroWhite	FullCure 705	Objet 30	28 μm	[38]

Table 1. Cont.

Main Material	Support Material	Machine	Layer Thickness	Reference
FLX935, VeroMagenta	SUP706B	Objet J750	-	[27]
VeroClear, VeroWhitePlus	FullCure 705	Objet 260 Connex2	-	[55]
VeroWhitePlus	FullCure 705	Objet 30	16 μm	[70]
-	-	Objet Eden 260VS	16 μm	[71]
VeroClear	SUP707	Objet Eden 260VS	-	[72]
FullCure 720	-	Objet Eden 260VS	16 μm	[73]
MED610	-	Objet Eden 260VS	-	[74]
RGD240	FullCure 705	Objet 30	28 μm	[36]
TangoBlackPlus VeroWhitePlus	-	Objet Connex 350	30 μm	[57]
FullCure 720	FullCure 705	Objet Eden 350	16 μm	[75]
RGD840	FullCure 705	Objet 30	28 μm	[76]
RGD515	-	Objet 350	16 μm	[77]
FullCure 720	-	Objet Eden 250	16 μm and 30 μm	[31,33]
RGD720	SUP706	Objet 30	-	[78]
Digital ABS Ivory, VeroGray, RGD720 and Rigur	-	Connex2 Objet 500	-	[56]
VeroClear	-	Objet 30 Prime	16–28 μm	[79]
VeroBlackPlus	SUP706B	Objet J750	27 μm	[50]
VeroWhite, RDG525, MED610	-	Objet 260 Connex1	-	[32]
Agilus30 and VeroWhite	SUP706	Objet 500 Connex3	-	[80]
VeroWhitePlus	FullCure 705	Objet 260 Connex	-	[48]
VeroClear	-	Objet 30 Prime	28 μm	[44]
FullCure 720	-	Objet Eden 250	16, 30 μm	[30]
FullCure 720	-	Objet Eden 250	16, 30 μm	[34]
RGD240	FullCure 705	Object 30	28 μm	[51]
FullCure 720, VeroWhite	FullCure 705	Objet Connex 350	32 μm	[81]
VeroWhitePlus	-	Object 30	16 μm	[82]

2.5. Comparisons with Other Technologies

Polymeric materials can be manufactured using different techniques in addition to MJ, as follows: selective laser sintering (SLS), fused deposition modeling (FDM or fused filament fabrication (FFF)), three-dimensional printing (3DP), binder jetting (BJ), stereolithography (SLA), color-jet printing (CJP), digital light processing (DLP) and laminated object manufacturing (LOM). These technologies were investigated and compared in different aspects in the literature.

Ramola et al. stated that MJ technology produces 3D models with higher accuracy than SLS or 3DP [83]. Ibrahim et al. investigated the dimensional accuracy of MJ, 3DP and SLS printed mandibular anatomy parts, which showed that SLS parts had a lower dimensional error (1.79%) than MJ (2.14%) and 3DP (3.14%) parts but MJ printed parts had the highest accuracy [84]. Salmi et al. compared dimensional accuracy of SLS, 3DP and MJ printed skull models and stated that the MJ method resulted in the lowest dimensional error (0.18% ± 0.12% for first measurement and 0.18% ± 0.13% for second measurement) [85].

MJ technology also gives rise to less dimensional variance than SLS technology due to the laser dispersion in the build plate [86].

It was asserted in the literature that MJ technology has higher resolution than FDM which may result in the existence of coarse weld lines between successive layers when FDM is used [87]. Lee et al. compared the surface roughness of replica teeth produced by FDM and MJ technology. They stated that MJ resulted in smoother surfaces than FDM due to lower layer thickness values (MJ: 0.016 mm, FDM: 0.330 mm) [82]. Camardella et al. investigated the dimensional accuracy of dental models made with SLA and MJ techniques and reported that all the models manufactured by MJ were accurate [73]. Maurya et al. compared FDM and MJ technology in terms of form error (flatness, roundness and cylindricity), dimensional accuracy, surface roughness, tolerance grade and cost analysis for an automotive part (engine connecting rod). They recorded that MJ printed parts have a lower percentage error along the XY plane, lower average percentage error in circular dimensions, lower form error and lower surface roughness but higher cost than FDM printed parts [76].

Kim et al. compared SLA, FDM, MJ, SLS, 3DP and LOM technologies in terms of dimensional accuracy, mechanical properties, surface roughness, printing speed and cost. They noted that MJ technology is advantageous in terms of tensile strength at room temperature [88]. Manoharan et al. compared SLA, MJ, SLS and 3DP in terms of surface finish, dimensional accuracy, materials and printing time for the production of sports footwear. They concluded that MJ showed the best dimensional accuracy, good surface finish, reasonable supported materials and short printing time [89]. Li et al. compared FDM, SLA and MJ technologies in terms of surface roughness, part cost, sustainability and human perception of surface texture and material colors. They concluded that MJ printed parts had the lowest surface roughness but the highest environmental impact and cost [90]. Queral et al. compared dimensional accuracy of SLS, SLA, FDM and MJ manufactured coil winding structures. They stated that high-quality MJ and FDM achieved 0.1% minimum dimensional accuracy for these parts [91]. Tan et al. compared three different AM techniques (MJ, FDM and SLS) in terms of dimensional accuracy. They showed that MJ resulted in the most accurate final part, whereas FDM resulted in the greatest dimensional deviation from the requirements [92].

One of the ongoing problems for additive manufacturing that needs to be overcome is the dimensional accuracy. In the comparison studies, it was found that MJ technology gives higher dimensional accuracy when printing dental models than SLA, DLP and FFF techniques [71]. Hong et al. compared three different AM techniques in terms of dimensional accuracy of thyroid cancer phantom parts: FDM, CJP and MJ. They stated that MJ gave the best results in terms of dimensional accuracy and clinical demands, but its cost was relatively high [93]. Chen et al. compared reproducibility, dimensional accuracy and dimensional stability of surgical templates produced by three different AM techniques: SLA, MJ and direct metal printing. They concluded that MJ printed templates had the highest accuracy and reproducibility, but their accuracy deteriorated after 1 month of storage [74]. Khaledi et al. compared three different production techniques in metal copings pattern fabrication: milling, SLA and MJ. They stated that the MJ method has a smaller marginal discrepancy than the SLA and milling techniques, meaning that MJ gives the highest accuracy [94]. Dizon et al. compared dimensional accuracy of injected parts from polymer molds produced by SLA, MJ and FDM technologies. They stated that the surface finish of SLA and MJ printed parts was excellent and dimensional accuracy of injected parts from MJ printed molds was higher than from SLA printed molds [77]. Msallem et al. compared the dimensional accuracies of anatomic mandibular models printed with five different AM methods: SLS, BJ, FFF, MJ and SLA. They stated that overall trueness analyses were carried out for SLS, BJ, FFF, MJ and SLA with decreasing trueness [88]. Park et al. compared dimensional variation in dental casts produced by FDM, DLP, SLA and MJ. They stated that casts printed by FDM and DLP showed contraction behavior, whereas casts printed by MJ and SLA showed expansion behavior [95]. Eliasova et al. investigated four different AM techniques in terms of surface roughness and dimensional accuracy. They concluded that MJ samples had fibrous structures and showed higher surface roughness as compared to other techniques. They also specified that surface roughness over the build direction is much higher than that over the perpendicular one for MJ samples (Figure 3) [96]. Budzik et al. compared car mirror holder parts manufactured by MJ, FDM and DLP in terms of visual control, application of a caliper and application of a contactless optical system. They stated that the MJ method was the most precise technique [97]. Wesemann et al. compared the accuracy of occlusal splints produced by SLA, DLP and MJ and concluded that SLA showed the smallest manufacturing deviations [98].

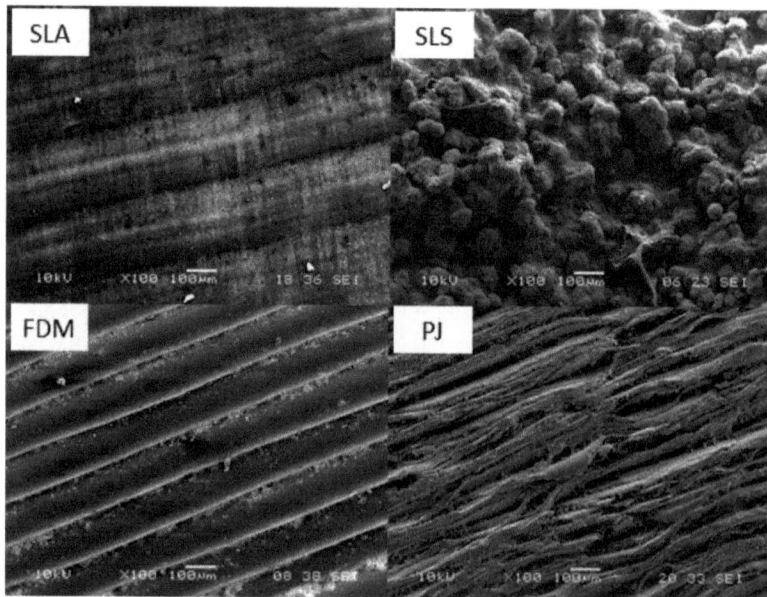

Figure 3. Printed surfaces for SLA, SLS, FDM and MJ samples [96].

3. Applications

3.1. Medical Applications

Due to the aforementioned advantages, MJ technology is utilized in different industries for different purposes. In the biomedical industry, MJ is exploited for printing hand protheses by using FullCure 720 material [99]. Manufacturing surgical training models [100] and printing multi-color bone models can also be achieved, which can potentially be used as anatomy teaching tools for specific diseases [101]. For example, printing human nasal sinus anatomy was demonstrated to teach different patients about their medical conditions and surgical treatment options (Figure 4) [102]. In one of the studies, Khalid et al. compared mechanical properties of three different types of materials in MJ: VeroWhitePlus (RGD835), TangoBlackPlus (FLX980) and RigidLightGrey25 (RGD8510-DM) and stated that these materials can be employed in pediatric head impact scenario investigations [103].

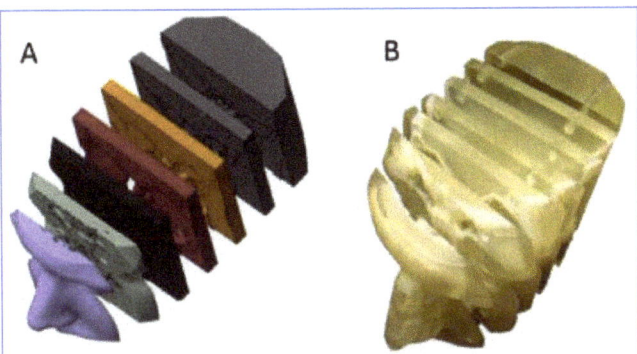

Figure 4. (**A**) CT scan slices of patient nasal cavity, (**B**) MJ printed model [102].

In dentistry, MJ is used for printing implant surgical templates, mouthpiece fixation instruments and implant guide production. Kim et al. produced maxilla and mandible implant guides with PolyJet, SLA and MultiJet printing (MJP) and stated that PolyJet parts had the lowest dimensional variations [63]. Herschdorfer et al. used PolyJet, SLA and MJP to produce implant surgical templates and stated that no significant effect was found in the accuracy of templates between the three AM processes [64]. Kitamori et al. printed mouthpiece fixation instruments used by head and neck radiotherapy patients by MJ. They replicated the mouthpieces by applying computed tomography to skull bones and produced them by MJ with MED610 material and stated that the templates can successfully be implemented in dosimetry [104]. Anunmana et al. produced implant guides with MJ, DLP and SLA and stated that MJ printed samples showed the highest 3D dimensional accuracy at entry point and apex [65]. Etajuri et al. produced implant guides by applying computed tomography to sheep mandibles with MJ. They stated that the dimensional variations are within acceptable limits of 2 mm [105].

3.2. Mechanical Applications

In the manufacturing industry, MJ is employed to manufacture plastic injection molds (Figure 5) [66,106] provide final parts ready to validate a product [107] and in tooling. Since MJ inserts have a smoother surface finish than direct metal laser sintering (DMLS) inserts, they can be utilized without any polishing [108].

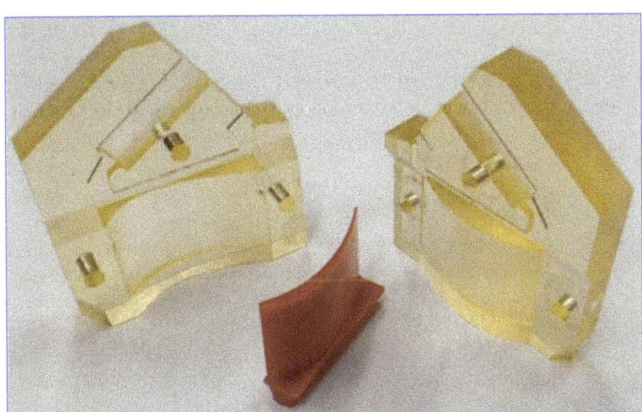

Figure 5. Wax turbine blade from MJ printed molds [106].

MJ was used to produce intricate 3D pentamode structures from FullCure850 VeroGray with minimum and maximum ligament thicknesses of 0.71 mm and 1.32 mm, respectively [109]. Honeycomb, re-entrant and auxetic structures was also produced from FullCure850 VeroGray material with MJ technology for crashbox applications [110]. MJ was used to produce dual-material auxetic metamaterials (DMAMs) to increase the mechanical properties and stability of metamaterials. For this purpose, DMAMs were produced by MJ with an accuracy of 0.1 mm from stiff and ductile materials and it was stated that the deformation behavior and Poisson's ratio of DMAMs in the elastic region were controlled and buckling in stiff regions was prevented [111]. MJ was also used in biomedical applications for control and actuation of robots and robot arms by using metamaterials from rubber-like materials. In one of the studies, it was stated that an inchworm-type soft robot for crawling through channels was produced by MJ and showed good stability and resilience [112].

3.3. Acoustic Applications

MJ technology is capable of producing parts with different materials and fillers. One of the applications based on multi-material and filler usage in MJ is producing parts by adding ceramic and metal materials to increase the acoustic ability of the metamaterials [113–115]. In one of the studies, special cellular thin-walled specimens based on resonant-type coupled tubes with open-end faces were produced by MJ from VeroWhite 830 material for acoustic applications where the frequency band ranged from 300 to 600 Hz and it was stated that MJ printed parts showed two times higher acoustic power absorption capability than a standard absorber of the same size where a mineral wool layer was covered by a perforated panel [116]. In another study, a sandwich panel with an MJ printed polymer core, unidirectional carbon-fiber composite front and back face sheets and translucent epoxy DP 190 to provide bonding between core and sheets were used. It was stated that this sandwich panel configuration increased the damping performance by nearly 25% without any optimization and it was also proposed that this configuration was a good candidate for pressurized airplane fuselage load-carrying flexural structures [117].

3.4. Electronics Applications

In electronics applications, MJ is also utilized due to its unique properties and ability to modify MJ printers to obtain needed structures. Jabari et al. developed and characterized a drop-on-demand piezoelectric–pneumatic material jetting (PPMJ) additive manufacturing process for the aim of printing graphene-based nanocomposites. The developed method exhibited a production rate 10 times faster than current extrusion techniques, in which printing speed reaches about 500 mm/s. As well as the production rate, this technology provides low resistive parts that can compete with other previously reported extrusion-based graphene structure printing methods [118]. Zhang et al. developed a polyimide film for MJ to create an insulating bridge for a circuit. As a result, the produced polyimide film showed a permittivity level of 3.41 and around 500 °C degradation temperature, which are comparable values to commercially polyimide films [119]. Moreover, MJ technology was exploited in the production of low-voltage polymer field-effect transistors [120], and all-polymer capacitors [121].

3.5. Multi-Material Applications

Multi-material additive manufacturing is a promising research topic that provides both geometrical complexity and material flexibility in one structure. Multi-material production with MJ requires additional print heads to disperse different materials for curing. Local changes in the material in a structure cause increased tensile strength, biocompability, flexibility and visuality with different colors and opacities. Thus, beyond aesthetic uses, a multi-material product is generally used to increase the energy absorption capabilities of lattice structures. Wang et al. produced multi-material auxetic lattice structures for increasing the energy absorption of a structure. Stiff materials were utilized to produce ligaments that are parallel to the compression direction; however, joints in which premature failures occur were fabricated using flexible materials [111]. Moreover, a similar method was also applied to another auxetic lattice structure with the same purpose [122]. As a result, the multi-material production capability of MJ provides design flexibility for energy absorption of structures.

3.6. Other Applications

In the aviation industry, MJ can be used in rapid manufacturing and be used to test various wing prototype designs for unmanned aerial vehicle (UAV) applications [4] and to produce wing structures with different types of lattice designs to achieve lightweight aircrafts [123].

MJ technology appeared in fashion industries for its multi-material production ability. For instance, a highly textured cape and skirt in Iris van Herpen's VOLTAGE collection were MJ printed [124]. Moreover, as an unexpected application area, musical instrument components [67] were manufactured via MJ, and biocompatible plastic toys [68] were also produced. Other interesting purposes for MJ technology include production of bellow actuators [125,126], microfluid mixing devices [127] and devices with integrated porous structures which are used for colorimetric detection of iron in natural waters and soil [69]. In addition, the technology was also adapted to design and fabricate acoustic metamaterial samples with high sound absorption efficiency. Vdovin et al. compared the sound absorption capability of standard absorbers and MJ printed absorbers. They used an Objet Eden 350 PolyJet machine and FullCure 720 and VeroWhite 830 materials. Their results revealed that MJ printed absorbers have a higher sound absorption capability than standard ones due to 3D printing without distortion of the intended geometry. Additionally, additional UV polymerization after the build process reduced residual stresses in the printed samples [116].

Additive manufacturing is associated with the production of complex arbitrary structures such as lattice structures. MJ technology is also utilized in the production of lattice geometries to be used in different types of testing instead of selective laser melting (SLM) technology which has been intensively used in lattice specimen production but the cost of which is much higher than for polymer-based parts [128]. In one study, MJ technology was used to evaluate the compressive behavior of triply periodic minimum surface lattice structures [129]. To overcome the limitation, the technology was used to produced special honeycomb structures in which a solid side wall of the traditional honeycomb structure can be replaced by a porous side wall, which resulted in improved stiffness and strength [130]. Composite polymer parts including ferromagnetic reinforcing particles can also be achieved by developing MJ technology [131].

4. Summary

This review focused on state of art of MJ printing technology, and its advantages over other technologies and applications. Today, MJ technology is used in very different industries due to its advantages in producing parts in a relatively short period of time. A detailed literature review showed that MJ technology results in higher dimensional accuracy and better surface roughness values than other polymeric material printing technologies (FDM, SLA, etc.). However, more improvements can still be made by changing the control variables. As consistently reported in the literature, tray location, post-processing, material type, layer thickness, surface finish and build orientation have considerable effects on mechanical properties, surface roughness and dimensional accuracy of the final part. For surface roughness, it was revealed that the surface finish setting has the most effect, whereas for dimensional accuracy; however, layer thickness and model scale are the secondary factors that affect the surface roughness. For better surface roughness values, glossy finish settings need to be chosen during printing. Studies also showed that parts' distribution on the tray affects mechanical properties and for higher strength results, part spacing needs to be decreased. Build orientation also has considerable influence on the characteristics of the final part and it was stated in the literature that when build orientation increases, surface roughness increases and dimensional accuracy and strength decrease. In the design approach, multi-material prints can be achieved for better or arbitrary mechanical performances. Additionally, obtaining different colors in one build enables better visual designs for different fields in which human perception is necessary.

Based on the findings in the literature, the following research gaps were identified:

- Industrial applications of MJ technology still need further investigation. Especially in the aviation industry, MJ was used only in small prototype wing applications. Its potential for aircraft modification purposes where mock-ups are needed for proper installation is worth investigating.
- MJ technology produces high-quality parts but for higher quality applications, some post-processing techniques need to be applied. Post-processing of MJ printed parts needs further investigation.
- Design engineers need some guidelines for proper designs to be manufactured. For this reason, designs for MJ manufacturing need further investigation.
- Dimensional variations in any production method are very important for proper installation. The producibility of MJ printed parts has been evaluated in the literature but, especially for parts used in aviation and automotive industries, producibility and dimensional variations still needs more attention.

Author Contributions: Conceptualization, O.G. and K.G.; methodology, O.G. and K.G.; validation, O.G. and K.G.; investigation, O.G. and K.G.; resources, O.G., K.G. and A.T.; writing—original draft preparation, O.G., K.G. and A.T.; writing—review and editing, O.G., K.G. and A.T. All authors have read and agreed to the published version of the manuscript.

Funding: This research was funded by the Technological and Scientific Council of Turkey (TUBITAK) Technology and Innovation Support Program, grant number 5158001.

Institutional Review Board Statement: Not applicable.

Informed Consent Statement: Not applicable.

Data Availability Statement: The data presented in this study are available on request from the corresponding author.

Conflicts of Interest: The authors declare no conflict of interest. The funders had no role in the design of the study; in the collection, analyses, or interpretation of data; in the writing of the manuscript, or in the decision to publish the results.

References

1. Flores, I.; Kretzschmar, N.; Azman, A.H.; Chekurov, S.; Pedersen, D.B.; Chaudhuri, A. Implications of lattice structures on economics and productivity of metal powder bed fusion. *Addit. Manuf.* **2020**, *31*, 100947. [CrossRef]
2. Avila, J.D.; Bose, S.; Bandyopadhyay, A. Additive manufacturing of titanium and titanium alloys for biomedical applications. In *Titanium in Medical and Dental Applications*; Froes, F.H., Qian, M., Eds.; Woodhead Publishing: Sawston, UK, 2018; pp. 325–343.
3. Herzog, D.; Seyda, V.; Wycisk, E.; Emmelmann, C. Additive manufacturing of metals. *Acta Mater.* **2016**, *117*, 371–392. [CrossRef]
4. Najmon, J.C.; Raeisi, S.; Tovar, A. Review of additive manufacturing technologies and applications in the aerospace industry. In *Additive Manufacturing for the Aerospace Industry*; Froes, F., Boyer, R., Eds.; Elsevier Publishing: Amsterdam, The Netherlands, 2019; pp. 7–31.
5. Udroiu, R.; Braga, I.C. PolyJet technology applications for rapid tooling. *MATEC Web Conf.* **2017**, *112*, 03011. [CrossRef]
6. ISO/ASTM 52900: 2015. Additive Manufacturing—General Principles—Terminology. Available online: https://www.iso.org/standard/69669.html (accessed on 25 June 2021).
7. Leary, M. *Design for Additive Manufacturing*; Elsevier Publishing: Amsterdam, The Netherlands, 2020; p. 270.
8. Pilipović, A.; Baršić, G.; Katić, M.; Havstad, M.R. Repeatability and reproducibility assessment of a PolyJet technology using x-ray computed tomography. *Appl. Sci.* **2020**, *10*, 7040. [CrossRef]
9. Quan, H.; Zhang, T.; Xu, H.; Luo, S.; Nie, J.; Zhu, X. Photo-curing 3D printing technique and its challenges. *Bioact. Mater.* **2020**, *5*, 110–115. [CrossRef]
10. Bagheri, A.; Jin, J. Photopolymerization in 3D printing. *ACS Appl. Polym. Mater.* **2019**, *1*, 593–611. [CrossRef]
11. Bennett, J. Measuring UV curing parameters of commercial photopolymers used in additive manufacturing. *Addit. Manuf.* **2017**, *18*, 203–212. [CrossRef]
12. Gardan, J. Additive manufacturing technologies: State of the art and trends. *Int. J. Prod. Res.* **2016**, *54*, 3118–3132. [CrossRef]

13. O'Neill, P.; Jolivet, L.; Kent, N.J.; Brabazon, D. Physical integrity of 3D printed parts for use as embossing tools. *Adv. Mater. Process. Technol.* **2017**, *3*, 308–317. [CrossRef]
14. Sireesha, M.; Lee, J.; Kiran, A.S.K.; Babu, V.J.; Kee, B.B.T.; Ramakrishna, S. A review on additive manufacturing and its way into the oil and gas industry. *RSC Adv.* **2018**, *8*, 22460–22468. [CrossRef]
15. Dilag, J.; Chen, T.; Li, S.; Bateman, S.A. Design and direct additive manufacturing of three-dimensional surface micro-structures using material jetting technologies. *Addit. Manuf.* **2019**, *27*, 167–174. [CrossRef]
16. Pilipović, A.; Raos, P.; Šercer, M. Experimental analysis of properties of materials for rapid prototyping. *Int. J. Adv. Manuf. Technol.* **2009**, *40*, 105–115. [CrossRef]
17. Kaweesa, D.V.; Meisel, N.A. Quantifying fatigue property changes in material jetted parts due to functionally graded material interface design. *Addit. Manuf.* **2018**, *21*, 141–149. [CrossRef]
18. Mirzaali, M.J.; Edens, M.E.; de la Nava, A.H.; Janbaz, S.; Vena, P.; Doubrovski, E.L.; Zadpoor, A.A. Length-scale dependency of biomimetic hard-soft composites. *Sci. Rep.* **2018**, *8*, 1–8. [CrossRef] [PubMed]
19. Ganesan, S.; Ranganathan, R. Design and development of customised split insole using additive manufacturing technique. *Int. J. Rapid Manuf.* **2018**, *7*, 295–309. [CrossRef]
20. Gu, G.X.; Takaffoli, M.; Hsieh, A.J.; Buehler, M.J. Biomimetic additive manufactured polymer composites for improved impact resistance. *Extrem. Mech. Lett.* **2016**, *9*, 317–323. [CrossRef]
21. Yap, Y.L.; Yeong, W.Y. Shape recovery effect of 3D printed polymeric honeycomb: This paper studies the elastic behaviour of different honeycomb structures produced by PolyJet technology. *Virtual Phys. Prototyp.* **2015**, *10*, 91–99. [CrossRef]
22. Li, W.; Xu, K.; Li, H.; Jia, H.; Liu, X.; Xie, J. Energy absorption and deformation mechanism of lotus-type porous coppers in perpendicular direction. *J. Mater. Sci. Technol.* **2017**, *33*, 1353–1361. [CrossRef]
23. Gu, G.X.; Takaffoli, M.; Buehler, M.J. Hierarchically enhanced impact resistance of bioinspired composites. *Adv. Mater.* **2017**, *29*, 1700060. [CrossRef]
24. Sugavaneswaran, M.; Arumaikkannu, G. Modelling for randomly oriented multi material additive manufacturing component and its fabrication. *Mater. Des.* **2014**, *54*, 779–785. [CrossRef]
25. Sugavaneswaran, M.; Arumaikkannu, G. Analytical and experimental investigation on elastic modulus of reinforced additive manufactured structure. *Mater. Des.* **2015**, *66*, 29–36. [CrossRef]
26. Dalaq, A.S.; Abueidda, D.W.; Al-Rub, R.K.A. Mechanical properties of 3D printed interpenetrating phase composites with novel architectured 3D solid-sheet reinforcements. *Compos. Part A Appl. Sci. Manuf.* **2016**, *84*, 266–280. [CrossRef]
27. Tee, Y.L.; Tran, P.; Leary, M.; Pille, P.; Brandt, M. 3D Printing of polymer composites with material jetting: Mechanical and fractographic analysis. *Addit. Manuf.* **2020**, *36*, 101558. [CrossRef]
28. Stansbury, J.W.; Idacavage, M.J. 3D printing with polymers: Challenges among expanding options and opportunities. *Dent. Mater.* **2016**, *32*, 54–64. [CrossRef]
29. Cazón, A.; Morer, P.; Matey, L. PolyJet technology for product prototyping: Tensile strength and surface roughness properties. *Proc. Inst. Mech. Eng. Part B J. Eng. Manuf.* **2014**, *228*, 1664–1675. [CrossRef]
30. Kechagias, J.; Iakovakis, V.; Giorgo, E.; Stavropoulos, P.; Koutsomichalis, A.; Vaxevanidis, N.M. Surface roughness optimization of prototypes produced by PolyJet direct 3D printing technology. In Proceedings of the International Conference on Engineering and Applied Sciences Optimization, Kos Island, Greece, 4–6 June 2014.
31. Aslani, K.-E.; Vakouftsi, F.; Kechagias, J.D.; Mastorakis, N.E. Surface roughness optimization of poly-jet 3D printing using Grey Taguchi method. In Proceedings of the 2019 International Conference on Control, Artificial Intelligence, Robotics & Optimization (ICCAIRO), Athens, Greece, 8–10 December 2019; IEEE: New York, NY, USA, 2019; pp. 213–218.
32. Kent, N.J.; Jolivet, L.; O'Neill, P.; Brabazon, D. An evaluation of components manufactured from a range of materials, fabricated using PolyJet technology. *Adv. Mater. Process. Technol.* **2017**, *3*, 318–329. [CrossRef]
33. Kechagias, J.; Stavropoulos, P.; Koutsomichalis, A.; Ntintakis, I.; Vaxevanidis, N. Dimensional accuracy optimization of prototypes produced by PolyJet direct 3D printing technology. In Proceedings of the International Conference on Industrial Engineering, Santorini Island, Greece, 18–20 July 2014; pp. 61–65.
34. Aslani, K.-E.; Korlos, A.; Kechagias, J.D.; Salonitis, K. Impact of process parameters on dimensional accuracy of PolyJet 3D printed parts using grey Taguchi method. In Proceedings of the MATEC Web of Conferences; EDP Sciences, 2020; Volume 318, p. 1015. Available online: https://www.matec-conferences.org/articles/matecconf/abs/2020/14/matecconf_icmmen20_01015/matecconf_icmmen20_01015.html (accessed on 9 July 2021).
35. Pugalendhi, A.; Ranganathan, R.; Gopalakrishnan, B. Effects of process parameters on build time of PolyJet printed parts using Taguchi method. In Proceedings of the International Conference on Advances in Materials Processing & Manufacturing Applications, National Institute of Technology (MNIT), Jaipur, India, 5–6 November 2020.
36. Gay, P.; Blanco, D.; Pelayo, F.; Noriega, A.; Fernández, P. Analysis of factors influencing the mechanical properties of flat PolyJet manufactured parts. *Procedia Eng.* **2015**, *132*, 70–77. [CrossRef]
37. Barclift, M.W.; Williams, C.B. Examining variability in the mechanical properties of parts manufactured via PolyJet direct 3D printing. In Proceedings of the International Solid Freeform Fabrication Symposium, Austin, TX, USA, 10–12 August 2012.
38. Beltrán, N.; Carriles, F.; Álvarez, B.J.; Blanco, D.; Rico, J.C. Characterization of factors influencing dimensional and geometric errors in PolyJet manufacturing of cylindrical features. *Procedia Eng.* **2015**, *132*, 62–69. [CrossRef]

39. Kumar, K.; Kumar, G.S. A study on surface roughness of rapid prototypes facricated using PolyJet 3D printing system. In Proceedings of the International Conference on Computer Aided Engineering (CAE 2013); 2013; pp. 1–6. Available online: https://www.researchgate.net/profile/Saravana-Kumar-Gurunathan/publication/294090024_A_STUDY_ON_SURFACE_ROUGHNESS_OF_RAPID_PROTOTYPES_FABRICATED_USING_POLY-JET_3D_PRINTING_SYSTEM/links/56be23390 8aeedba05610e31/A-STUDY-ON-SURFACE-ROUGHNESS-OF-RAPID-PROTOTYPES-FABRICATED-USING-POLY-JET-3D-PRINTING-SYSTEM.pdf (accessed on 10 July 2021).
40. Kumar, K.; Kumar, G.S. An experimental and theoretical investigation of surface roughness of poly-jet printed parts: This paper explains how local surface orientation affects surface roughness in a poly-jet process. *Virtual Phys. Prototyp.* **2015**, *10*, 23–34. [CrossRef]
41. Kechagias, J.D.; Maropoulos, S. An investigation of sloped surface roughness of direct poly-jet 3D printing. In *Proceedings of the Proceedings of the International Conference on Industrial Engineering—INDE*; 2015; pp. 150–153. Available online: http://universitypress.org.uk/library/2015/zakynthos/bypaper/CIMC/CIMC-26.pdf (accessed on 9 July 2021).
42. Khoshkhoo, A.; Carrano, A.L.; Blersch, D.M. Effect of surface slope and build orientation on surface finish and dimensional accuracy in material jetting processes. *Procedia Manuf.* **2018**, *26*, 720–730. [CrossRef]
43. Vidakis, N.; Petousis, M.; Vaxevanidis, N.; Kechagias, J. Surface roughness investigation of Poly-Jet 3D printing. *Mathematics* **2020**, *8*, 1758. [CrossRef]
44. Haghighi, A.; Yang, Y.; Li, L. Dimensional performance of as-built assemblies in polyjet additive manufacturing process. In Proceedings of the International Manufacturing Science and Engineering Conference, Los Angeles, CA, USA, 4–8 June 2017; American Society of Mechanical Engineers: New York, NY, USA, 2017; Volume 50732, p. V002T01A039.
45. Kitsakis, K.; Kechagias, J.; Vaxevanidis, N.; Giagkopoulos, D. Tolerance Analysis of 3d-MJM parts according to IT grade. In Proceedings of the IOP Conference Series: Materials Science and Engineering, Kozani, Greece, 23–25 September 2016; IOP Publishing: Bristol, UK, 2016; Volume 161, p. 12024.
46. Kitsakis, K.; Kechagias, J.; Vaxevanidis, N.; Giagkopoulos, D. Tolerance assessment of PolyJet direct 3D printing process employing the IT grade approach. *Acad. J. Manuf. Eng.* **2016**, *14*, 62–68. [CrossRef]
47. Kesy, A.; Kotlinski, J. Mechanical properties of parts produced by using polymer jetting technology. *Arch. Civ. Mech. Eng.* **2010**, *10*, 37–50. [CrossRef]
48. Das, S.C.; Ranganathan, R.; Murugan, N. Effect of build orientation on the strength and cost of PolyJet 3D printed parts. *Rapid Prototyp. J.* **2018**, *24*, 832–839. [CrossRef]
49. Tomar, P.R.S.; Ulu, F.I.; Kelkar, A.; Mohan, R.V. Investigation of process induced variations in PolyJet printing with digital polypropylene via homogeneous 3D tensile test coupon. In Proceedings of the ASME 2019 International Mechanical Engineering Congress and Exposition, Salt Lake City, UT, USA, 11–14 November 2019.
50. Sanders, J.; Wei, X.; Pei, Z. *Experimental Investigation of Stratasys J750 PolyJet Printer: Effects of Orientation and Layer Thickness on Thermal Glass Transition Temperature*; International Mechanical Engineering Congress and Exposition: Salt Lake City, UT, USA, 2019.
51. Blanco, D.; Fernandez, P.; Noriega, A. Nonisotropic experimental characterization of the relaxation modulus for PolyJet manufactured parts. *J. Mater. Res.* **2014**, *29*, 1876–1882. [CrossRef]
52. Reichl, K.K.; Inman, D.J. Dynamic mechanical and thermal analyses of Objet connex 3D printed materials. *Exp. Tech.* **2018**, *42*, 19–25. [CrossRef]
53. Yap, Y.L.; Wang, C.C.; Sing, S.L.; Dikshit, V.; Yeong, W.Y.; Wei, J. Material jetting additive manufacturing: An experimental study using designed metrological benchmarks. *Precis. Eng.* **2017**, *50*, 275–285. [CrossRef]
54. Pugalendhi, A.; Ranganathan, R.; Chandrasekaran, M. Effect of process parameters on mechanical properties of VeroBlue material and their optimal selection in PolyJet technology. *Int. J. Adv. Manuf. Technol.* **2020**, *108*, 1049–1059. [CrossRef]
55. Pugalendhi, A.; Ranganathan, R.; Ganesan, S. Impact of process parameters on mechanical behaviour in multi-material jetting. *Mater. Today Proc.* **2020**. [CrossRef]
56. Kampker, A.; Kreisköther, K.; Reinders, C. Material and parameter analysis of the PolyJet process for mold making using design of experiments. *Int. J. Mater. Metall. Eng.* **2017**, *11*, 242–249. [CrossRef]
57. Moore, J.P.; Williams, C.B. Fatigue properties of parts printed by PolyJet material jetting. *Rapid Prototyp. J.* **2015**, *21*, 675–685. [CrossRef]
58. Udroiu, R.; Braga, I.C. System performance and process capability in additive manufacturing: Quality control for polymer jetting. *Polymers* **2020**, *12*, 1292. [CrossRef] [PubMed]
59. Silva, M.R.; Pereira, A.M.; Sampaio, Á.M.; Pontes, A.J. Assessment of the Dimensional and Geometric Precision of Micro-Details Produced by Material Jetting. *Materials* **2021**, *14*, 1989. [CrossRef]
60. Baršic, G.; Pilipovic, A.; Katic, M. Reproducibility of 3D printed structures. In Proceedings of the Proc. of Euspen's 17th Int. Conf. & Exhibition, Hannover, ICE17144; 2017.
61. Udroiu, R.; Braga, I.C.; Nedelcu, A. Evaluating the quality surface performance of additive manufacturing systems: Methodology and a material jetting case study. *Materials* **2019**, *12*, 995. [CrossRef] [PubMed]
62. Dangnan, F.; Espejo, C.; Liskiewicz, T.; Gester, M.; Neville, A. Friction and wear of additive manufactured polymers in dry contact. *J. Manuf. Process.* **2020**, *59*, 238–247. [CrossRef]
63. Kim, T.; Lee, S.; Kim, G.B.; Hong, D.; Kwon, J.; Park, J.W.; Kim, N. Accuracy of a simplified 3D-printed implant surgical guide. *J. Prosth. Dent.* **2020**, *124*, 195–201. [CrossRef]

64. Herschdorfer, L.; Negreiros, W.M.; Gallucci, G.O.; Hamilton, A. Comparison of the accuracy of implants placed with CAD-CAM surgical templates manufactured with various 3D printers: An in vitro study. *J. Prosth. Dent.* **2020**, *125*, 905–910. [CrossRef]
65. Anunmana, C.; Ueawitthayasuporn, C.; Kiattavorncharoen, S.; Thanasrisuebwong, P. In vitro comparison of surgical implant placement accuracy using guides fabricated by three different additive technologies. *Appl. Sci.* **2020**, *10*, 7791. [CrossRef]
66. Singh, R. Comparison of PolyJet printing and silicon moulding as rapid plastic moulding solutions. *Int. J. Automot. Mech. Eng.* **2012**, *6*, 777–784. [CrossRef]
67. Cottrell, S.; Howell, J. Reproducing musical instrument components from manufacturers' technical drawings using 3D printing: Boosey & Hawkes as a case study. *J. New Music Res.* **2019**, *48*, 449–457. [CrossRef]
68. Martínez-García, A.; Sandoval-Pérez, I.; Ibáñez-García, A.; Pernías-Peco, K.; Varela-Gandía, F.J.; Galvañ-Gisbert, J. Influence of process parameters of different additive manufacturing techniques on mechanical properties and safety of customised toys. *Procedia Manuf.* **2019**, *41*, 106–113. [CrossRef]
69. Balavandy, S.K.; Li, F.; Macdonald, N.P.; Maya, F.; Townsend, A.T.; Frederick, K.; Guijt, R.M.; Breadmore, M.C. Scalable 3D printing method for the manufacture of single-material fluidic devices with integrated filter for point of collection colourimetric analysis. *Anal. Chim. Acta* **2020**, *1151*, 238101. [CrossRef]
70. Bhat, M.A.; Shaikh, A.A. Effect of specimen parameters on mixed-mode I/II stress intensity factors for additive manufactured slant edge crack plate. *Mater. Today Proc.* **2021**, *44*, 4305–4308. [CrossRef]
71. Kim, S.-Y.; Shin, Y.-S.; Jung, H.-D.; Hwang, C.-J.; Baik, H.-S.; Cha, J.-Y. Precision and trueness of dental models manufactured with different 3-dimensional printing techniques. *Am. J. Orthod. Dentofac. Orthop.* **2018**, *153*, 144–153. [CrossRef]
72. Gupta, V.; Mahbub, P.; Nesterenko, P.N.; Paull, B. A new 3D printed radial flow-cell for chemiluminescence detection: Application in ion chromatographic determination of hydrogen peroxide in urine and coffee extracts. *Anal. Chim. Acta* **2018**, *16*, 81–92. [CrossRef] [PubMed]
73. Camardella, L.T.; Vilella, O.V.; Breuning, H. Accuracy of printed dental models made with 2 prototype technologies and different designs of model bases. *Am. J. Orthod. Dentofac. Orthop.* **2017**, *151*, 1178–1187. [CrossRef] [PubMed]
74. Chen, L.; Lin, W.-S.; Polido, W.D.; Eckert, G.J.; Morton, D. Accuracy, reproducibility, and dimensional stability of additively manufactured surgical templates. *J. Prosthet. Dent.* **2019**, *122*, 309–314. [CrossRef] [PubMed]
75. Udroiu, R.; Nedelcu, A.; Deaky, B. Rapid manufacturing by PolyJet technology of customized turbines for renewable energy generation. *Environ. Eng. Manag. J.* **2011**, *10*, 1387–1394. [CrossRef]
76. Maurya, N.K.; Rastogi, V.; Singh, P. Comparative study and measurement of form errors for the component printed by FDM and PolyJet process. *J. Homepage Httpiieta Orgjournalsi2m* **2019**, *18*, 353–359. [CrossRef]
77. Dizon, J.R.C.; Valino, A.D.; Souza, L.R.; Espera, A.H.; Chen, Q.; Advincula, R.C. 3D printed injection molds using various 3D printing technologies. In Proceedings of the Materials Science Forum; Trans Tech Publ: Baech, Switzerland, 2020; Volume 1005, pp. 150–156. Available online: https://www.scientific.net/MSF.1005.150 (accessed on 12 July 2021).
78. Nowacki, J.; Sieczkiewicz, N. Problems of determination of MultiJet 3D printing distortions using a 3D scanner. *Arch. Mater. Sci. Eng.* **2020**, *103*, 30–41. [CrossRef]
79. Emir, F.; Ayyıldız, S. Accuracy evaluation of complete-arch models manufactured by three different 3D printing technologies: A three-dimensional analysis. *J. Prosthodont. Res.* **2020**. [CrossRef]
80. Childs, E.H.; Latchman, A.V.; Lamont, A.C.; Hubbard, J.D.; Sochol, R.D. Additive assembly for PolyJet-based multi-material 3D printed microfluidics. *J. Microelectromech. Syst.* **2020**, *29*, 1094–1096. [CrossRef]
81. Kozior, T.; Kundera, C. Rheological properties of cellular structures manufactured by additive PJM technology. *Teh. Vjesn.* **2021**, *28*, 82–87. [CrossRef]
82. Lee, K.Y.; Cho, J.W.; Chang, N.Y.; Chae, J.M.; Kang, K.H.; Kim, S.C.; Cho, J.H. Accuracy of three-dimensional printing for manufacturing replica teeth. *Korean J. Orthod.* **2015**, *45*, 217–225. [CrossRef] [PubMed]
83. Ramola, M.; Yadav, V.; Jain, R. On the adoption of additive manufacturing in healthcare: A literature review. *J. Manuf. Technol. Manag.* **2019**, *30*, 48–69. [CrossRef]
84. Ibrahim, D.; De oliveira, M.G.; Nobre, S.M.W.; Silva, D.N. Dimensional error of selective laser sintering, three-dimensional printing and PolyJetTM models in the reproduction of mandibular anatomy. *J. Cranio-Maxillofac. Sur.* **2009**, *37*, 167–173. [CrossRef] [PubMed]
85. Salmi, M.; Paloheimo, K.; Tuomi, J.; Wolff, J.; Mäkitie, A. Accuracy of medical models made by additive manufacturing (rapid manufacturing). *J. Cranio-Maxillofac. Sur.* **2013**, *41*, 603–609. [CrossRef]
86. Paz, R.; Monzón, M.D.; Benítez, A.N.; González, B. New lightweight optimisation method applied in parts made by selective laser sintering and PolyJet technologies. *Int. J. Comput. Integr. Manuf.* **2016**, *29*, 462–472. [CrossRef]
87. Hosseinabadi, H.G.; Bagheri, R.; Gray, L.A.; Altstädt, V.; Drechsler, K. Plasticity in polymeric honeycombs made by photo-polymerization and nozzle based 3D-printing. *Polym. Test.* **2017**, *63*, 163–167. [CrossRef]
88. Kim, G.D.; Oh, Y.T. A benchmark study on rapid prototyping processes and machines: Quantitative comparisons of mechanical properties, accuracy, roughness, speed, and material cost. *Proc. Inst. Mech. Eng. Part B J. Eng. Manuf.* **2008**, *222*, 201–215. [CrossRef]
89. Manoharan, V.; Chou, S.M.; Forrester, S.; Chai, G.B.; Kong, P.W. Application of additive manufacturing techniques in sports footwear: This paper suggests a five-point scoring technique to evaluate the performance of four AM techniques, namely, stereolithography (SLA), PolyJet (PJ), selective laser sintering (SLS) and t. *Virtual Phys. Prototyp.* **2013**, *8*, 249–252. [CrossRef]

90. Li, Y.; Linke, B.S.; Voet, H.; Falk, B.; Schmitt, R.; Lam, M. Cost, sustainability and surface roughness quality–A comprehensive analysis of products made with personal 3D printers. *CIRP J. Manuf. Sci. Technol.* **2017**, *16*, 1–11. [CrossRef]
91. Queral, V.; Rincón, E.; Mirones, V.; Rios, L.; Cabrera, S. Dimensional accuracy of additively manufactured structures for modular coil windings of stellarators. *Fusion Eng. Des.* **2017**, *124*, 173–178. [CrossRef]
92. Tan, W.S.; Suwarno, S.R.; An, J.; Chua, C.K.; Fane, A.G.; Chong, T.H. Comparison of solid, liquid and powder forms of 3D printing techniques in membrane spacer fabrication. *J. Membr. Sci.* **2017**, *537*, 283–296. [CrossRef]
93. Hong, D.; Lee, S.; Kim, T.; Baek, J.H.; Lee, Y.; Chung, K.-W.; Sung, T.-Y.; Kim, N. Development of a personalized and realistic educational thyroid cancer phantom based on CT images: An evaluation of accuracy between three different 3D printers. *Comput. Biol. Med.* **2019**, *113*, 103393. [CrossRef]
94. Khaledi, A.A.; Farzin, M.; Akhlaghian, M.; Pardis, S.; Mir, S. Evaluation of the marginal fit of metal copings fabricated by using 3 different CAD-CAM techniques: Milling, stereolithography, and 3D wax printer. *J. Prosthet. Dent.* **2020**, *124*, 81–86. [CrossRef]
95. Park, J.; Jeon, J.; Koak, J.; Kim, S.; Heo, S. Dimensional accuracy and surface characteristics of 3D-printed dental casts. *J. Prosthet. Dent.* **2020**. [CrossRef] [PubMed]
96. Eliasova, H.; Dostalova, T.; Jelinek, M.; Remsa, J.; Bradna, P.; Prochazka, A.; Kloubcova, M. Surface morphology of three-dimensionally printed replicas of upper dental arches. *Appl. Sci.* **2020**, *10*, 5708. [CrossRef]
97. Budzik, G.; Woźniak, J.; Paszkiewicz, A.; Przeszłowski, Ł.; Dziubek, T.; Dębski, M. Methodology for the quality control process of additive manufacturing products made of polymer materials. *Materials* **2021**, *14*, 2202. [CrossRef] [PubMed]
98. Wesemann, C.; Spies, B.C.; Schaefer, D.; Adali, U.; Beuer, F.; Pieralli, S. Accuracy and its impact on fit of injection molded, milled and additively manufactured occlusal splints. *J. Mech. Behav. Biomed. Mater.* **2021**, *114*, 104179. [CrossRef] [PubMed]
99. Kate, J.T.; Smit, G.; Breedveld, P. 3D-printed upper limb prostheses: A review. *Disabil. Rehabil. Assist. Technol.* **2017**, *12*, 300–314. [CrossRef]
100. Hanisch, M.; Kroeger, E.; Dekiff, M.; Timme, M.; Kleinheinz, J.; Dirksen, D. 3D-printed surgical training model based on real patient situations for dental education. *Int. J. Environ. Res. Public Health* **2020**, *17*, 2901. [CrossRef] [PubMed]
101. Inoue, M.; Freel, T.; Avermaete, A.V.; Leevy, W.M. Color enhancement strategies for 3D printing of x-ray computed tomography bone data for advanced anatomy teaching models. *Appl. Sci.* **2020**, *10*, 1571. [CrossRef]
102. Sander, I.M.; Liepert, T.T.; Doney, E.L.; Leevy, W.M.; Liepert, D.R. Patient education for endoscopic sinus surgery: Preliminary experience using 3D-printed clinical imaging data. *J. Funct. Biomater.* **2017**, *8*, 13. [CrossRef] [PubMed]
103. Khalid, G.A.; Bakhtiarydavijani, H.; Whittington, W.R.; Prabhu, R.; Jones, M.D. Material response characterization of three poly jet printed materials used in a high fidelity human infant skull. *Mater. Today Proc.* **2020**, *20*, 408–413. [CrossRef]
104. Kitamori, H.; Sumida, I.; Tsujimoto, T.; Shimamoto, H.; Murakami, S.; Ohki, M. Evaluation of mouthpiece fixation devices for head and neck radiotherapy patients fabricated in PolyJet photopolymer by a 3D printer. *Phys. Med.* **2019**, *58*, 90–98. [CrossRef]
105. Etajuri, E.A.; Suliman, E.; Mahmood, W.A.A.; Ibrahim, N.; Buzayan, M.; Mohd, N.R. Deviation of dental implants placed using a novel 3D-printed surgical guide: An in vitro study. *Dent. Med. Probl.* **2020**, *57*, 359–362. [CrossRef]
106. Turek, P.; Budzik, G.; Sęp, J.; Oleksy, M.; Józwik, J.; Przeszłowski, Ł.; Paszkiewicz, A.; Kochmański, Ł.; Żelechowski, D. An analysis of the casting polymer mold wear manufactured using PolyJet method based on the measurement of the surface topography. *Polymers* **2020**, *12*, 3029. [CrossRef] [PubMed]
107. León-Cabezas, M.A.; Martínez-García, A.; Varela-Gandía, F.J. Innovative advances in additive manufactured moulds for short plastic injection series. *Procedia Manuf.* **2017**, *13*, 732–737. [CrossRef]
108. Mendible, G.A.; Rulander, J.A.; Johnston, S.P. Comparative study of rapid and conventional tooling for plastics injection molding. *Rapid Prototyp. J.* **2017**, *23*, 344–352. [CrossRef]
109. Schittny, R.; Bückmann, T.; Kadic, M.; Wegener, M. Elastic measurements on macroscopic three-dimensional pentamode metamaterials. *Appl. Phys. Lett.* **2013**, *103*, 231905. [CrossRef]
110. Simpson, J.; Kazancı, Z. Crushing investigation of crash boxes filled with honeycomb and re-entrant (auxetic) lattices. *Thin-Walled Struct.* **2020**, *150*, 106676. [CrossRef]
111. Wang, K.; Chang, Y.-H.; Chen, Y.; Zhang, C.; Wang, B. Designable dual-material auxetic metamaterials using three-dimensional printing. *Mater. Des.* **2015**, *67*, 159–164. [CrossRef]
112. Mark, A.G.; Palagi, S.; Qiu, T.; Fischer, P. Auxetic metamaterial simplifies soft robot design. In Proceedings of the 2016 IEEE international conference on robotics and automation (ICRA), Stockholm, Sweden, 16–21 May 2016; IEEE: New York, NY, USA, 2016; pp. 4951–4956.
113. Derby, B.; Reis, N. Inkjet printing of highly loaded particulate suspensions. *MRS Bull.* **2003**, *28*, 815–818. [CrossRef]
114. Ko, S.H.; Chung, J.; Hotz, N.; Nam, K.H.; Grigoropoulos, C.P. Metal nanoparticle direct inkjet printing for low-temperature 3D micro metal structure fabrication. *J. Micromechanics Microengineering* **2010**, *20*, 125010. [CrossRef]
115. Lucklum, F.; Vellekoop, M.J. Design and fabrication challenges for millimeter-scale three-dimensional phononic crystals. *Crystals* **2017**, *7*, 348. [CrossRef]
116. Vdovin, R.; Tomilina, T.; Smelov, V.; Laktionova, M. Implementation of the additive PolyJet technology to the development and fabricating the samples of the acoustic metamaterials. *Procedia Eng.* **2017**, *176*, 595–599. [CrossRef]
117. Yu, T.; Lesieutre, G.A. Damping of sandwich panels via three-dimensional manufactured multimode metamaterial core. *AIAA J.* **2017**, *55*, 1440–1449. [CrossRef]

118. Jabari, E.; Liravi, F.; Davoodi, E.; Lin, L.; Toyserkani, E. High speed 3D material-jetting additive manufacturing of viscous Graphene-based ink with high electrical conductivity. *Addit. Manuf.* **2020**, *35*, 101330. [CrossRef]
119. Zhang, F.; Saleh, E.; Vaithilingam, J.; Li, Y.; Tuck, C.J.; Hague, R.J.M.; Wildman, R.D.; He, Y. Reactive material jetting of polyimide insulators for complex circuit board design. *Addit. Manuf.* **2019**, *25*, 477–484. [CrossRef]
120. Liu, Y.; Varahramyan, K.; Cui, T. Low-voltage all-polymer field-effect transistor fabricated using an inkjet printing technique. *Macromol. Rapid Commun.* **2005**, *26*, 1955–1959. [CrossRef]
121. Liu, Y.; Cui, T.; Varahramyan, K. All-polymer capacitor fabricated with inkjet printing technique. *Solid. State. Electron.* **2003**, *47*, 1543–1548. [CrossRef]
122. Mirzaali, M.J.; Caracciolo, A.; Pahlavani, H.; Janbaz, S.; Vergani, L.; Zadpoor, A.A. Multi-material 3D printed mechanical metamaterials: Rational design of elastic properties through spatial distribution of hard and soft phases. *Appl. Phys. Lett.* **2018**, *113*, 241903. [CrossRef]
123. Goh, G.D.; Agarwala, S.; Goh, G.L.; Dikshit, V.; Yeong, W.Y. Additive manufacturing in unmanned aerial vehicles (UAVs): Challenges and potential. *Aerosp. Sci. Technol.* **2017**, *63*, 140–151. [CrossRef]
124. Vanderploeg, A.; Lee, S.; Mamp, M. The application of 3D printing technology in the fashion industry. *Int. J. Fash. Des. Technol. Educ.* **2017**, *10*, 170–179. [CrossRef]
125. Dämmer, G.; Gablenz, S.; Hildebrandt, A.; Major, Z. Design and shape optimization of PolyJet bellows actuators. In Proceedings of the 2018 IEEE International Conference on Soft Robotics (RoboSoft), Livorno, Italy, 24–28 April 2018; IEEE: New York, NY, USA, 2018; pp. 282–287.
126. Dämmer, G.; Gablenz, S.; Hildebrandt, A.; Major, Z. PolyJet-printed bellows actuators: Design, structural optimization, and experimental investigation. *Front. Rob. AI* **2019**, *6*, 34. [CrossRef]
127. Zeraatkar, M.; Filippini, D.; Percoco, G. On the impact of the fabrication method on the performance of 3D printed mixers. *Micromachines* **2019**, *10*, 298. [CrossRef]
128. Lancea, C.; Campbell, I.; Chicos, L.-A.; Zaharia, S.-M. Compressive behaviour of lattice structures manufactured by PolyJet technologies. *Polymers* **2020**, *12*, 2767. [CrossRef] [PubMed]
129. Sathishkumar, N.; Vivekanandan, N.; Balamurugan, L.; Arunkumar, N.; Ahamed, I. Mechanical properties of triply periodic minimal surface based lattices made by PolyJet printing. *Mater. Today Proc.* **2020**, *22*, 2934–2940. [CrossRef]
130. Liu, Y.; Zhou, M.; Fu, K.; Yu, M.; Zheng, G. Optimal design, analysis and additive manufacturing for two-level stochastic honeycomb structure. *Int. J. Comput. Integr. Manuf.* **2019**, *32*, 682–694. [CrossRef]
131. Aguilera, A.F.E.; Nagarajan, B.; Fleck, B.A.; Qureshi, A.J. Ferromagnetic particle structuring in material jetting-Manufacturing control system and software development. *Procedia Manuf.* **2019**, *34*, 545–551. [CrossRef]

Article

Development and Characterization of Field Structured Magnetic Composites

Balakrishnan Nagarajan [1], Yingnan Wang [1], Maryam Taheri [2,3], Simon Trudel [2], Steven Bryant [3], Ahmed Jawad Qureshi [1] and Pierre Mertiny [1,*]

[1] Department of Mechanical Engineering, University of Alberta, Edmonton, AB T6G 1H9, Canada; bnagaraj@ualberta.ca (B.N.); yingnan1@ualberta.ca (Y.W.); ajqureshi@ualberta.ca (A.J.Q.)
[2] Department of Chemistry, University of Calgary, Calgary, AB T2N 1N4, Canada; maryam.taheri2@ucalgary.ca (M.T.); trudels@ucalgary.ca (S.T.)
[3] Department of Petroleum and Chemical Engineering, University of Calgary, Calgary, AB T2N 1N4, Canada; steven.bryant@ucalgary.ca
* Correspondence: pmertiny@ualberta.ca

Abstract: Polymer composites containing ferromagnetic fillers are promising for applications relating to electrical and electronic devices. In this research, the authors modified an ultraviolet light (UV) curable prepolymer to additionally cure upon heating and validated a permanent magnet-based particle alignment system toward fabricating anisotropic magnetic composites. The developed dual-cure acrylate-based resin, reinforced with ferromagnetic fillers, was first tested for its ability to polymerize through UV and heat. Then, the magnetic alignment setup was used to orient magnetic particles in the dual-cure acrylate-based resin and a heat curable epoxy resin system in a polymer casting approach. The alignment setup was subsequently integrated with a material jetting 3D printer, and the dual-cure resin was dispensed and cured in-situ using UV, followed by thermal post-curing. The resulting magnetic composites were tested for their filler loading, microstructural morphology, alignment of the easy axis of magnetization, and degree of monomer conversion. Magnetic characterization was conducted using a vibrating sample magnetometer along the in-plane and out-of-plane directions to study anisotropic properties. This research establishes a methodology to combine magnetic field induced particle alignment along with a dual-cure resin to create anisotropic magnetic composites through polymer casting and additive manufacturing.

Keywords: magnetic polymer composites; anisotropic properties; dual-cure resin; polymer casting; additive manufacturing

Citation: Nagarajan, B.; Wang, Y.; Taheri, M.; Trudel, S.; Bryant, S.; Qureshi, A.J.; Mertiny, P. Development and Characterization of Field Structured Magnetic Composites. *Polymers* **2021**, *13*, 2843. https://doi.org/10.3390/polym13172843

Academic Editor: Swee Leong Sing

Received: 5 August 2021
Accepted: 19 August 2021
Published: 24 August 2021

Publisher's Note: MDPI stays neutral with regard to jurisdictional claims in published maps and institutional affiliations.

Copyright: © 2021 by the authors. Licensee MDPI, Basel, Switzerland. This article is an open access article distributed under the terms and conditions of the Creative Commons Attribution (CC BY) license (https://creativecommons.org/licenses/by/4.0/).

1. Introduction

Magnetic composites are fundamental elements in many electrical, electronic and electromagnetic devices. Mixing magnetic powders with a polymeric binder that bonds and insulates the powder grains enables manufacturing magnetic composites for a variety of applications. Advantages of magnetic composites include the ability to be produced with complicated shapes, high dimensional accuracy, and good mechanical, magnetic, and physical properties. Magnetic composites are divided into hard magnetic and soft magnetic composites [1]. Hard magnetic composites, also known as polymer bonded permanent magnets, find significant applications in motors, sensors and actuators in automation equipment and medical devices. Common methods utilized to manufacture bonded magnets include compression molding, injection molding, calendaring and extrusion [2–4]. The properties of magnetic composites depend on the powder particle size and shape, type of polymeric binder, and the manufacturing process utilized [5]. Field structured magnetic composites fabricated by applying an external magnetic field during the polymerization process exhibited high remanence due to local field effects. It was observed that a uniaxial field produced chain-like particles, and a biaxial field produced sheet-like particle structures in the fabricated composites [6]. Composites with magnetostrictive features were developed using an injection molding method where carbonyl iron particles were field structured in a thermoplastic elastomer matrix using an electromagnet. It was observed that samples with aligned iron particles exhibited a higher modulus compared to randomly dispersed magnetic polymer composites [7].

Functional magnetic composites, magnetic shape memory alloys, magnetic micro electromechanical systems (MEMS) and magnetic elastomers use a variety of magnetic materials [8]. Magnetic forces offer an attractive option for actuation in MEMS devices due to contact-free actuation capabilities. Microscale magnetic actuation capabilities have led to the implementation in a variety of microfluids and MEMS devices. In the field of micro-robotics, magnetic forces are used to provide wireless control and power to perform complex three-dimensional motions. Integration of different magnetic fillers with the polymer resin remains a significant challenge in the fabrication process [9]. Fabrication of micro magnets using traditional ultraviolet light (UV) lithography and micro-molding techniques have already been reported in the technical literature. However, technical challenges like adjusting suspension viscosity for spin-coating processes, particle settling, and precise control of particle alignment still exist in the fabrication of magnetically loaded polymer composites for microscale applications [10]. A micro pump with diffuser elements and an integrated composite magnet was developed using neodymium iron boron (NdFeB) magnet powder dispersed in polydimethylsiloxane (PDMS) resin [11]. High-performance NdFeB micro magnets using magnetron sputtering and high power plasma etching techniques have been reported in the technical literature [12]. Using low-modulus membrane materials, elastic hard magnetic films with the ability to produce bi-directional deflections in an external magnetic field were created using microfabrication approaches [13]. Mechanically compliant, magnetically responsive micro structures using a ferromagnetic photoresist containing nickel nanospheres dispersed in photosensitive epoxy resin (SU8) were fabricated using UV lithography-based approaches [14]. Screen printing was used as a technique to fabricate strontium ferrite thick films with an easy axis orientation perpendicular to the film surface [15]. Hard magnetic films were fabricated by embedding anisotropic NdFeB particles in a polymethylmethacrylate polymer matrix for MEMS applications. Fabricated thick films exhibited out of plane macroscopic magnetic anisotropy [16]. The immense capabilities of magnetic field responsive materials in terms of magnetic actuation, deformation capabilities like stretching, bending, and rotation upon exposure to magnetic fields, abilities for controllable drug release, and shape memory behavior have made polymer based magnetic materials a topic of intensive research [17]. Even though many robust manufacturing approaches exist, new innovations in materials and manufacturing processes can significantly enhance the performance of many functional devices. One such fast emerging technology is additive manufacturing, which will be detailed in the following.

Additive manufacturing (AM) is gaining momentum towards developing bonded magnets and magnetic composites with a wide range of magnetic particles and polymeric matrix materials. The fabrication of bonded magnets by adopting material jetting AM processes with epoxy resin based magnetic pastes was demonstrated in the technical literature [18–20]. Rheological additives and multimodal magnetic mixtures were utilized to develop the feedstock materials for material jetting AM processes. A multi extruder 3D printer was utilized by Yan et al. to fabricate magnetic components for power electronics applications [21]. Magnetic material formulations developed using UV curable resins were printed using nScrypt tabletop micro dispensing equipment that enabled the precise control of printed layer thickness of the deposited material [22,23]. Manufacturing of permanent magnets using extrusion-based AM of thermoplastic based feedstock materials has been detailed in the technical literature [24–27]. The ability of thermoplastic materials to be remelted under the influence of heat enabled conducting in situ magnetic alignment studies of NdFeB alloy powder reinforced ethylene vinyl acetate co-polymer-based permanent magnets. It was observed that the alignment temperature, magnetic field strength, and properties of the polymer matrix influenced the degree of particle alignment [28]. Anisotropic NdFeB/samarium ferrite nitride based bonded magnets were developed adopting a post printing alignment approach using a vibrating sample magnetometer, leading to an enhancement in magnetic properties in magnetic field aligned samples [29]. Similar magnetic field aligned bonded magnets were developed using anisotropic NdFeB powder reinforcing a polyamide matrix [30,31]. In-situ magnetic alignment using an electromagnet integrated with fused deposition modelling equipment was tested on anisotropic NdFeB/samarium ferrite nitride powders reinforcing a Nylon 12 matrix. It was found that the degree of alignment was a function of applied magnetic field and printing temperature [32]. It was additionally noted that alignment studies were primarily conducted in thermoplastic polymer-bonded permanent magnets.

In this research the authors modified a UV curable acrylate-based prepolymer for heat-initiated polymerization. An in-house developed permanent magnet-based particle alignment system aimed to orient magnetic particles within the polymer matrix was tested with the dual-cure acrylate-based resin and a commercially available epoxy prepolymer. The epoxy resin was included in the present study as a baseline due to its well-established material characteristics. Experiments were initially conducted through polymer casting where 3D printed molds were utilized to hold the liquid prepolymer. The alignment magnetic jig was further integrated with material jetting based 3D printing equipment, and the dual-cure acrylate-based prepolymer was used to 3D print magnetic composites. Scanning electron microscopy was employed to observe the distribution of magnetic particles within the polymer matrix. X-ray diffraction was utilized to evaluate the orientation of the easy axis of magnetization in the magnetic composites. Samples were tested for in-plane and out-of-plane magnetic properties using a vibrating sample magnetometer. Findings from this study are expected to advance the fabrication of anisotropic magnetic thermoset polymer composites for applications in sensors and other electrical and electronic devices.

2. Materials

For this research, anisotropic $Nd_2Fe_{14}B$ (herein abbreviated as NdFeB) powder of type MQA-38-14 was purchased from Magnequench Inc. (Pendleton, IN, USA). The UV curable urethane acrylate prepolymer PR48 was obtained from Colorado Photopolymer Solutions (Boulder, CO, USA). To induce heat-initiated free radical polymerization, the thermal initiator 2,2′-Azobis(2-methylpropionitrile), generally referred to as AIBN, was purchased from Sigma Aldrich (St. Louis, MO, USA). For the epoxy resin, EPON 826 with the aromatic amine curing agent Epicure W were procured from Hexion Inc. (Columbus, OH, USA). The permanent magnets used to construct the magnetic alignment array, N52 grade NdFeB magnets, were purchased from K&J Magnetics, Inc. (Pipersville, PA, USA).

3. Experimental Methods

3.1. Dual-Cure Resin Formulation Preparation and Testing

Exposing the magnetic particle reinforced prepolymer to UV, due to opacity, resulted in solidification of only the topmost surface and leaving the underlying material in an uncured state. This observation motivated the development of a formulation with the capability to cure under the influence of both UV and heat. A measured quantity of the thermal initiator was added to the UV curable prepolymer and mechanically stirred using an impeller. The resultant mixture was allowed to rest for a day prior to adding the magnetic filler to allow complete dissolution of the thermal initiator. The prepared prepolymer was transferred to small aluminum cuvettes and subsequently exposed to UV and further cured at approximately between 50 °C to 70 °C in a gravity convection oven.

3.2. Fabrication of Magnetic Field Structured Composites Using Polymer Casting and AM

In this study, the ability of a permanent magnet-based particle alignment fixture to orient magnetic particles in prepolymer formulations was tested. Two different material formulations namely, NdFeB powder with approximately 80 wt% dispersed in the dual-cure acrylate-based resin and the epoxy resin were utilized in this work. Small molds of dimensions 15 mm by 15 mm by 10 mm, 3D printed using a stereolithography printer, were utilized to hold the prepolymer formulation during the curing and alignment process. The oven temperature for curing the epoxy resin was between 60 °C to 80 °C.

To 3D print the dual-cure resin and develop anisotropic magnetic composites, a material jetting 3D printer integrated with a magnetic array-based particle alignment system was developed and utilized. The material jetting device was controlled using the Labview computing environment (National Instruments, Austin, TX, USA). The sample geometry, designed using Solidworks (Dassault Systems, Veliz-Villacoublay, France), was further processed using the open-source software Slic3r to generate the G-code for the nozzle deposition path. The generated G-code was further modified to accommodate particle alignment steps after every deposited layer.

3.3. Characterization of Magnetic Polymer Composites

Using scanning electron microscopy (SEM), the morphology of the magnetic particle reinforced composites was analyzed with a Zeiss Sigma 300 VP field emission scanning electron microscope (Oberkochen, Germany). The composite samples were cut, polished and coated with carbon. The latter was accomplished using a Leica EM SCD005 evaporative carbon coater (Wetzlar, Germany).

Thermogravimetric analysis (TGA) was used as a tool to evaluate the resultant magnetic filler loading in the manufactured magnetic polymer composites. TGA was conducted using a Discovery TGA (TA Instruments, Delaware, USA) from 25 °C to 600 °C in a nitrogen atmosphere at a flow rate of 20 mL/min and at a heating rate of 20 °C/min.

X-ray diffraction (XRD) analysis was utilized to characterize the orientation of the easy axis of magnetization in the manufactured magnetic composites. XRD measurements were performed using a Geigerflex diffractometer (Rigaku Corporation, Tokyo, Japan) fitted with a cobalt tube as an X-ray source and a graphite monochromator to filter the K_β wavelength. Tests were conducted at 38 kV and 38 mA, and the samples were scanned over 2θs ranging from 30° to 70° at a rate of 2°/min.

Fourier transform infrared spectroscopy (FTIR) analysis was conducted to evaluate the degree of monomer conversion in the composites fabricated using the dual-cure acrylate-based prepolymer. FTIR analysis was conducted using a Nicolet iS50 spectrometer (Thermo Fisher Scientific, Waaltham, MA, USA). FTIR spectra were collected over a range from 400 cm^{-1} to 4000 cm^{-1} at a resolution of 4 cm^{-1}. The uncured prepolymer was also characterized to compare its peak intensities with the cured polymers, which is indicative of the degree of monomer conversion.

The magnetic properties of fabricated composites were measured using a vibrating sample magnetometer (VSM, Versa Lab, Quantum Design, San Diego, CA, USA), in two conditions: parallel to the direction of field (in-plane) and perpendicular to the direction of field (out-of-plane). A small sample piece was sectioned using a handsaw, weighed and placed in the brass sample holder. For in-plane measurements, samples were secured between two quartz braces. For the out-of-plane measurement, a small amount of GE 7031 varnish and Kapton tape were also used. The effect of quartz, GE varnish and Kapton tape on the measurements is negligible. Magnetic hysteresis loops were derived by applying a magnetic field with a strength of $\mu_0 H = \pm 3$ Tesla at a temperature of 300 K. The magnetic properties were determined by averaging the values obtained through both magnetization and demagnetization cycles.

4. Results and Discussion

4.1. Evaluation of Dual-Cure Resin Formulation

UV-curable resin formulations require UV for the reaction initiation step. Free radicals are only generated during UV irradiation which enable reaction propagation and crosslinking of the monomers. One of the challenges with magnetic particle reinforced UV curable prepolymers is to achieve complete solidification as a result of polymerization. UV curable prepolymers reinforced with magnetic fillers experience lower levels of monomer conversion due to opacity introduced by the magnetic fillers. In addition to free radical generation through photochemical methods, a heat-initiated monomer conversion was introduced by adding the AIBN thermal initiator to the polymer formulation [33]. Initiator loading fractions ranging from 0.1 wt% to 0.4 wt% were tested. Adding AIBN at 0.4 wt% to a formulation containing magnetic fillers at 80 wt% was observed to render a cured magnetic polymer composite. Polymer casting was adopted for the initial trials. The prepolymer was observed to cure within an hour at temperatures ranging from 50 °C to 70 °C. A temperature range is indicated to account for any fluctuations within the oven. Initial results shown in Figure 1 indicate the ability of the material to fully cure through heat-initiated free radical polymerization. The cured composite was cut and smoothed using emery cloth for further observations (see photographs in Figure 2) exhibiting some voids in a solid material.

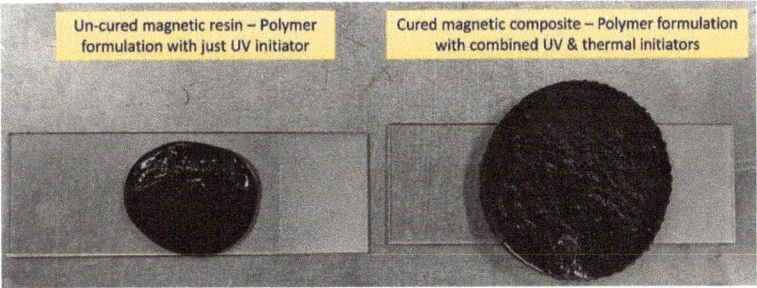

Figure 1. Thermal cure evaluation of dual-cure prepolymer reinforced with magnetic fillers.

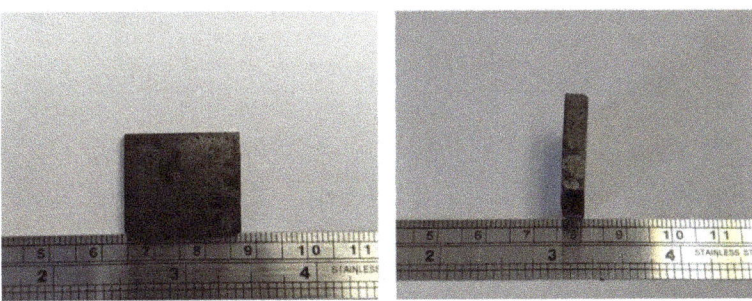

Figure 2. Cut and ground sample of cured magnetic composite created using dual-cure prepolymer reinforced with magnetic fillers.

4.2. Manufacturing Field Structured Magnetic Composites Using Polymer Casting and Additive Manufacturing

The magnetic alignment setup utilized to develop field structured magnetic composites was composed of a magnetic array with eight cube-shaped permanent magnets arranged in an orientation to deliver a uniaxial magnetic field. In a previous study, through the finite element method in magnetics, it was observed that the magnitude of magnetic flux density produced by the magnetic array was close to 0.3 Tesla [34]. The development of the uniaxial magnetic field within the magnetic array was further confirmed using optical microscopy where particle alignment within a magnetic particle reinforced resin droplet was evaluated [34]. The photograph in Figure 3 shows the setup prepared for manufacturing field structured magnetic composites. Securing the mold filled with the magnetic particle reinforced prepolymer was critical as the mold lost its stability within the magnetic array due to the magnetic forces. The fixture to hold the alignment magnets in the magnetic array, the mold securing fixture, and the resin holding mold were all 3D printed using the Autodesk Ember Digital Light Processing (DLP) 3D printer (Autodesk Inc., San Rafael, CA, USA). Using this method, magnetic composites with the dual-cure prepolymer and epoxy-based prepolymer were fabricated and tested.

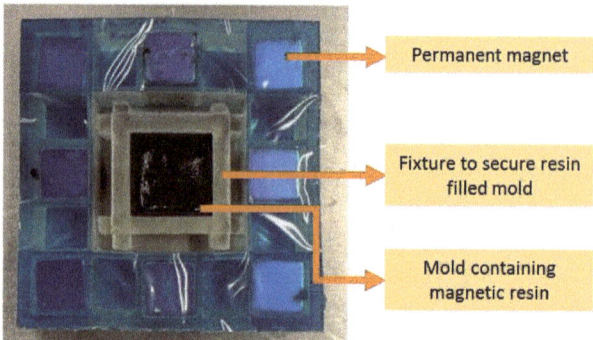

Figure 3. Magnetic alignment setup with permanent magnet array used to develop field structured magnetic composites.

Following the polymer casting trials, the dual-cure resin formulation was utilized to 3D print magnetic composites using the material jetting AM process. The process parameters of deposition pressure and speed were adjusted to 3 kPa and 10 mm/s, respectively. Thickness settings were 0.35 mm for the initially deposited layer and 0.5 mm for subsequent layer deposition. The photograph in Figure 4 shows a single deposited layer on an acrylic sheet substrate. Note that UV curing of the deposited layer in conjunction with particle alignment using the magnetic array was accomplished using 'manufacturing scenario B' established in a previous publication by the present authors [35]. In summary, the process step for each layer involves (i) material deposition, (ii) magnetic particle alignment for a set time, and (iii) UV exposure for 60 s. The photographs in Figure 5 show the material jetting equipment with the in-situ particle alignment system and the UV source for subsequent curing.

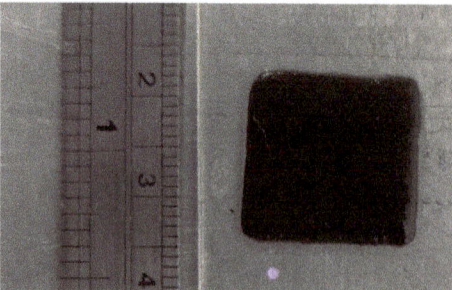

Figure 4. Single material layer deposited using the material-jetting 3D printer.

After completion of the thermal cure, the thickness of a single cured layer was measured using a digital caliper as shown in the photographs in Figure 6. Notably, with 0.84 mm the final thickness for the single-layer sample exceeded substantially the respective process parameter (0.35 mm). For multilayer samples the process of prepolymer deposition, particle alignment and UV curing was repeated, followed by thermal curing of samples. The photographs in Figure 7 show a 3D printed and cured sample consisting of three layers. The sample was cut and smoothed using emery cloth, leading to a slightly reduced sample thickness. Again, the measurement indicated in the figure reveals a final sample thickness that considerably exceeds the nominal thickness based on process parameters (i.e., 1.93 mm > 0.35 mm + 2 × 0.5 mm = 1.35 mm). Clearly, the used process parameters were not adjusted for print accuracy when using the filler modified prepolymer, which is considered a task beyond the scope of the work presented herein.

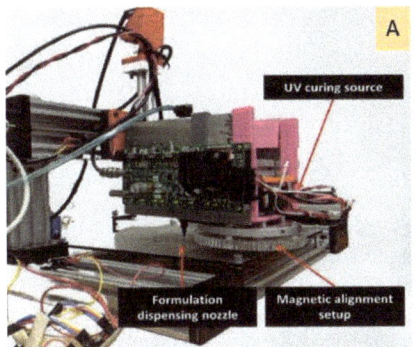
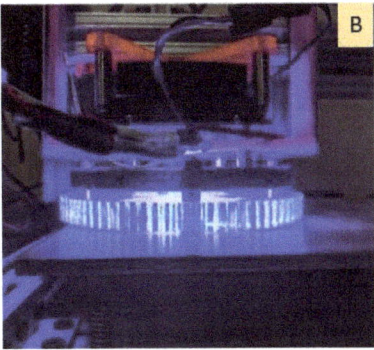

Figure 5. (**A**) Material jetting equipment with the in-situ particle alignment system and (**B**) UV source irradiating polymer sample for curing.

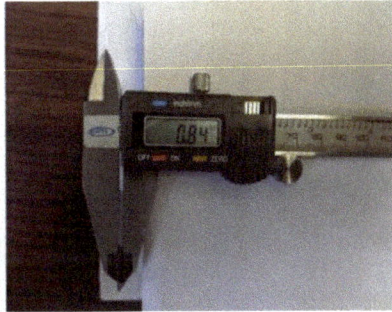

Figure 6. Appearance and dimensions of a single cured layer of 3D printed magnetic composite.

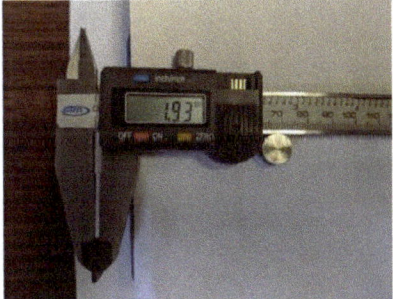

Figure 7. Appearance and dimensions of a cut and smoothed 3D printed magnetic composite with three print layers.

4.3. Characterization of Magnetic Polymer Composites

This section details the tests that were conducted to characterize the properties of fabricated magnetic composites (see Table 1), including composite morphology, filler loading, monomer conversion and magnetic anisotropy.

Table 1. List of fabricated and characterized sample types.

Sample	Material	Manufacturing Process	Magnetic Alignment
S1	MQA-38-14 dispersed in dual-cure prepolymer	Material-jetting AM	Yes
S2	MQA-38-14 dispersed in heat-curable epoxy resin	Polymer casting	Yes
S3	MQA-38-14 dispersed in dual-cure prepolymer	Polymer casting	Yes
S4	MQA-38-14 dispersed in dual-cure prepolymer	Polymer casting	No

4.3.1. Scanning Electron Microscopy

The morphology of composites consisting of magnetic particles in a polymer matrix was explored using SEM. The fundamental goal of the SEM study was to identify specific filler distribution and/or alignment characteristics due to the externally applied magnetic field. SEM micrographs of magnetic composites taken at two different magnifications are shown in Figures 8 and 9. In the samples manufactured via polymer casting (S2 and S3) coupled with the magnetic field source, the SEM images exhibit particles structured in a specific direction. Arrows included in Figure 9 indicate apparent particle stacking and alignment. Such striations irrespective of the particle size elucidate the influence of an externally applied magnetic field. Similar microstructural features for field structured magnetic composites were observed by Gandha et al. [31] for anisotropic NdFeB powder reinforced polyamide, where the magnetic filler was aligned using a post printing alignment field of 1 Tesla. Differences in the apparent degree of particle alignment and stacking between sample S2 and S3 in Figure 9 are likely related to the prepolymer viscosities. From previous research it is known that the viscosity of the acrylate-based prepolymer is approximately 0.3 Pa·s whereas the value for the epoxy resin is 7 Pa·s [20,35].

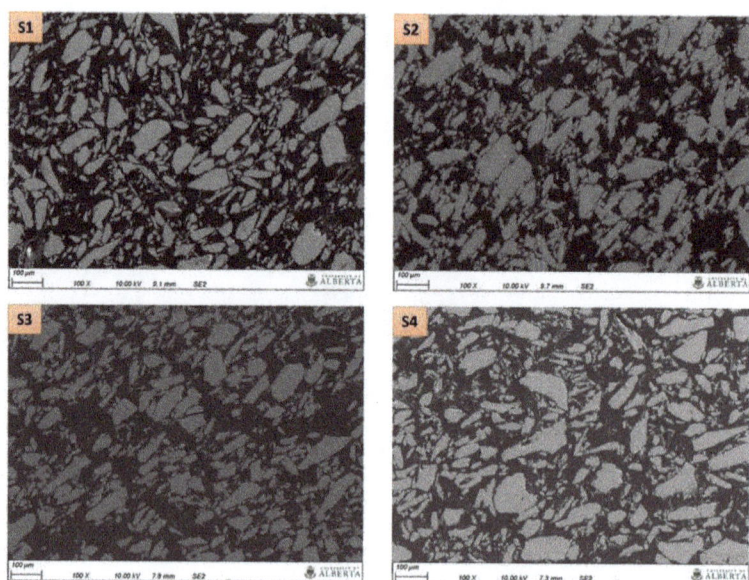

Figure 8. SEM images of magnetic composites as listed in Table 1.

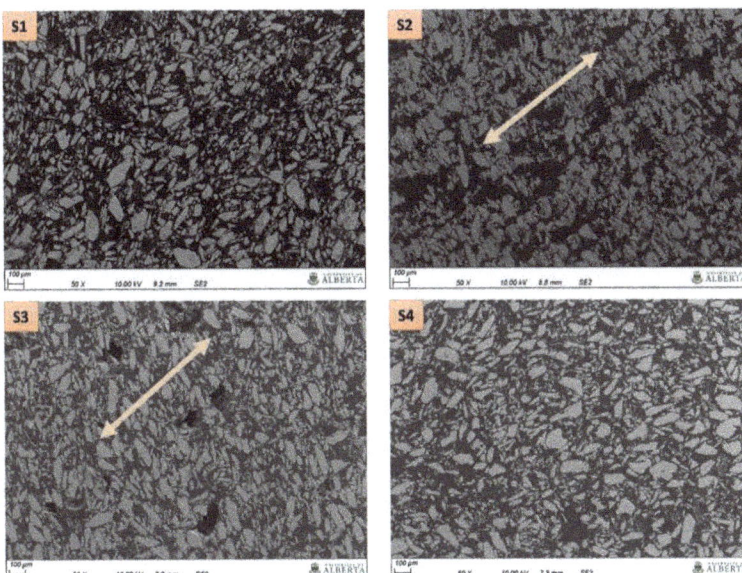

Figure 9. SEM micrographs of magnetic composites as listed in Table 1. Arrows indicate apparent particle stacking and alignment.

Alignment or stacking features are not apparent in Figure 9 for the sample fabricated without application of an external magnetic field source (S4), suggesting an isotropic composite morphology. Referring to sample S1 that was 3D printed using material jetting and an applied magnetic field source, many particles appear to exhibit structuring similar to samples S2 and S3, yet the microstructure also exhibits features similar to the isotropic magnetic composite. Given this ambiguity, it is difficult to confirm whether or not alignment or stacking features are due to the externally applied magnetic field or other effects (e.g., occupied volume effects). It should be mentioned that for the S1 sample the magnetic particle alignment time was shorter than for the samples produced by polymer casting. Adjusting the alignment time for 3D printed samples was necessary to mitigate material deformation under the influence of the magnetic field because no mold is providing material containment in this case (as discussed in [35]).

Besides morphological features relating to particles participles, resin rich spaces and porosities can be observed in the manufactured samples in Figures 8 and 9. These observations suggest that increased particle loading is feasible and that process adjustments need to be made to reduce porosity, with the latter being outside the scope of this paper.

4.3.2. Thermogravimetric Analysis

TGA was used to determine the magnetic particle loading in the fabricated magnetic composites. The thermal resin removal enabled determining the remaining weight of the sample in the crucible. The experimental data obtained from the TGA tests are depicted in Figure 10. The step transition temperature for samples S1, S2, S3, and S4 were observed to be 411 °C, 380 °C, 407 °C, and 413 °C, respectively. The percentage weight at 600 °C (end of the test) was taken as the weight percentage of the remaining magnetic filler material in the polymer composite. Table 2 lists the percentage weight of the residue at different temperatures obtained from TGA testing. It can be observed that the residue weight percentage of the magnetic filler exceeds the nominal filler weight fraction of the prepolymer formulation (80 wt%) by a maximum of about 5%. It is speculated that this discrepancy is rooted in material handling and manufacturing processes where effects such as material settling may have led to a slight increase in final filler weight fraction.

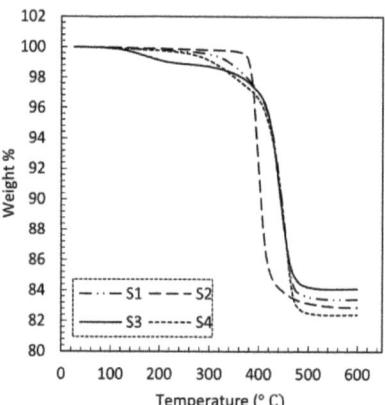

Figure 10. TGA data of magnetic composites as listed in Table 1.

Table 2. Residue weight percentage at different temperatures for composites as listed in Table 1.

Sample	Residue wt% at 200 °C	Residue wt% at 600 °C
S1	99.7	83.4
S2	99.8	82.9
S3	99.1	84.1
S4	99.7	82.4

4.3.3. X-ray Diffraction Analysis

X-ray diffraction was used to identify the orientation of the easy axis of magnetization in the composites as a result of the externally applied magnetic field. XRD analysis was conducted only for sample S2 and an additional isotropic sample fabricated using epoxy as the base prepolymer. In the presence of a magnetic field, a magnetic moment is generated along the easy axis of magnetization which results in particles interacting and forming aligned microstructures [35]. First, the magnetic powder was characterized to identify and confirm the chemical composition. The obtained XRD data was matched to the Powder Diffraction File reference data for NdFeB (number 00-039-0473). The 2θ incident angles representing the crystallographic c-axis were identified to facilitate comparisons for isotropic and anisotropic magnetic composites. Corresponding XRD data for 2θ incident angles ranging between 30° to 60° are depicted in Figure 11. From technical literature, it is understood that for NdFeB an enhancement of intensity of the (006) and (004) crystallographic plane indicates the orientation of the easy axis of magnetization in the magnetic composites [36]. Additionally, the disappearance of dominant peaks typically observed for a magnetic composite with an isotropic particle distribution is seen in Figure 11A. Results observed herein are congruent with the findings reported in the technical literature for field structured magnetic composites [31], and hence, an aligned microstructure can be ascertained for the S2 sample that was exposed to the magnetic field during manufacturing.

4.3.4. FTIR Spectroscopy

FTIR spectra of the developed dual-cure liquid prepolymer and the corresponding magnetic composite cured by UV and heat are shown in the graph in Figure 12. Note that the composite sample was in the form of a crushed powder. The obtained spectra were inspected primarily for reductions in peak intensities (i.e., peak flattening) that directly correlate to monomer conversion in the polymer composite [37,38]. Signals that correspond to the carbon double bond (C=C) in the acrylate along the regions 800 cm^{-1} to 830 cm^{-1} (=C-H out of plane bend), 1400 cm^{-1} to 1430 cm^{-1} (C=C twisting), and 1600 cm^{-1} to 1660 cm^{-1} (C=C stretching) were observed to diminish in the cured polymer composite compared to the uncured prepolymer, indicating that the composite was cured under the influence of heat in addition to UV irradiation [39].

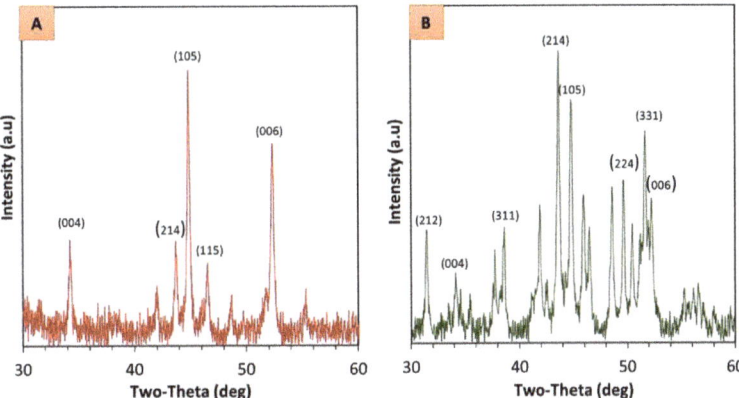

Figure 11. XRD patterns of (**A**) anisotropic magnetic composite fabricated with external magnetic field application, and (**B**) isotropic magnetic composite fabricated without an external magnetic field.

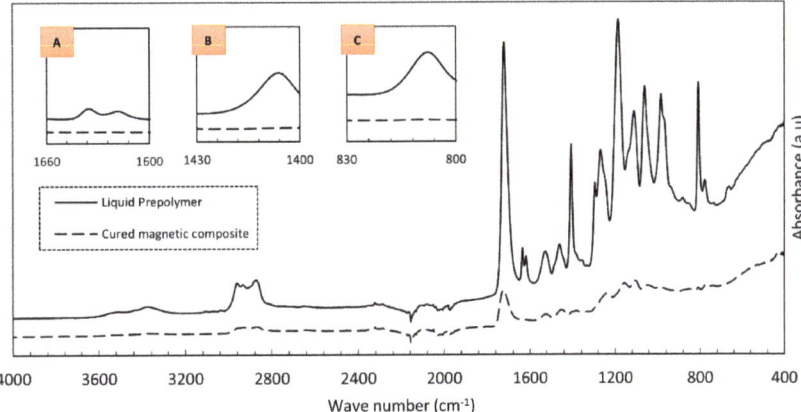

Figure 12. FTIR spectra of uncured liquid prepolymer and cured magnetic composite. Insets indicate FTIR spectral regions: (**A**) 1600 cm^{-1} to 1660 cm^{-1}, (**B**) 1400 cm^{-1} to 1430 cm^{-1}, and (**C**) 800 cm^{-1} to 830 cm^{-1}.

4.3.5. Magnetic Characterization

Composites reinforced with magnetic particles can be characterized by their magnetic saturation, remanence, and coercivity. Composites containing hard ferromagnetic particles should generate sufficient magnetic flux for a given application once they are magnetized, i.e., energized by applying a magnetic field using electromagnets. However, the properties of magnetic composites are dependent on microstructure, temperature, and demagnetizing fields [31]. The graphs in Figure 13 depict hysteresis loops measured using the vibrating sample magnetometer along the direction of particle structuring (in-plane) and perpendicular to the direction of particle structuring (out-of-plane). The samples fabricated applying an external magnetic field (S1, S2 and S3) exhibit significant enhancements in magnetic properties (saturation magnetization, squareness, and coercivity) along the in-plane easy direction over the out-of-plane direction. The presence of an applied magnetic field can be clearly seen to result in a preferred magnetization direction and strong anisotropy. This magnetic anisotropy is conferred by the preferential alignment of the magnetic filler powders along the curing applied field, as evidenced in Figures 8, 9 and 11. Field-structured magnetic composites developed using polymer casting (S2 and S3) exhibited anisotropic magnetic properties comparable to, or greater than, the isotropic specimen (S4). For the latter, no distinguishable differences can be ascertained between the in-plane and out-of-plane hysteresis loops. While the 3D printed sample also exhibits enhanced in-plane properties over the out-of-plane direction, it needs to be noticed that its properties are lower compared to the other samples. For example, the remanence to saturation ratio, which characterizes the degree of anisotropy, was observed to be approximately 0.85 for samples fabricated using polymer casting and 0.67 for the 3D printed sample. As mentioned earlier, a lower particle alignment time was employed for 3D printed samples (to mitigate magnetic resin deformation [35]), which is seen to have caused the somewhat reduced magnetic properties. Nevertheless, even with the reduced alignment time, the composite exhibited pronounced anisotropic and in-plane magnetic properties comparable to the other samples.

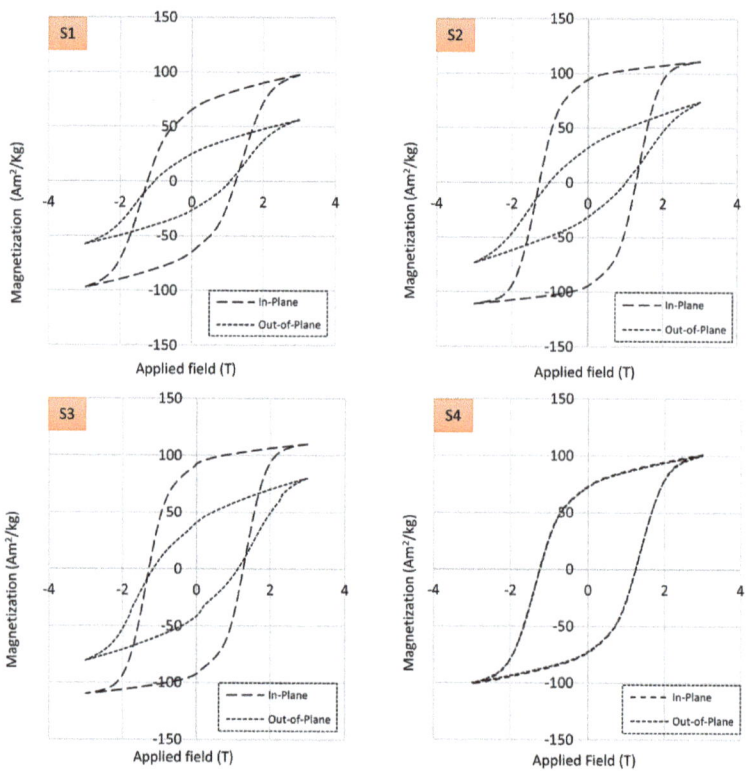

Figure 13. Hysteresis loops of magnetic composites as listed in Table 1, measured in sample in-plane and out-of-plane orientation using a vibrating sample magnetometer.

Data derived from magnetization and demagnetization cycles in the hysteresis graphs in Figure 13 are reduced to key properties in terms of magnetic saturation (at 3 Tesla), remanence, and coercivity in the bar graphs in Figure 14. The bar graphs indicate the efficacy of field structuring to achieve higher magnetic properties, i.e., saturation, remanence, and coercivity for the in-plane direction are consistently higher for samples S2 and S3 compared to the isotropic sample S4. Present observations are congruent to magnetic composites containing field structured particles in a thermoplastic matrix [31,32]. Overall, the present study validates the effectiveness of field structuring magnetic particles in thermoset polymers using a permanent magnet array, enabling an efficient fabrication of anisotropic magnetic composites for a variety of applications in electrical and electronic devices.

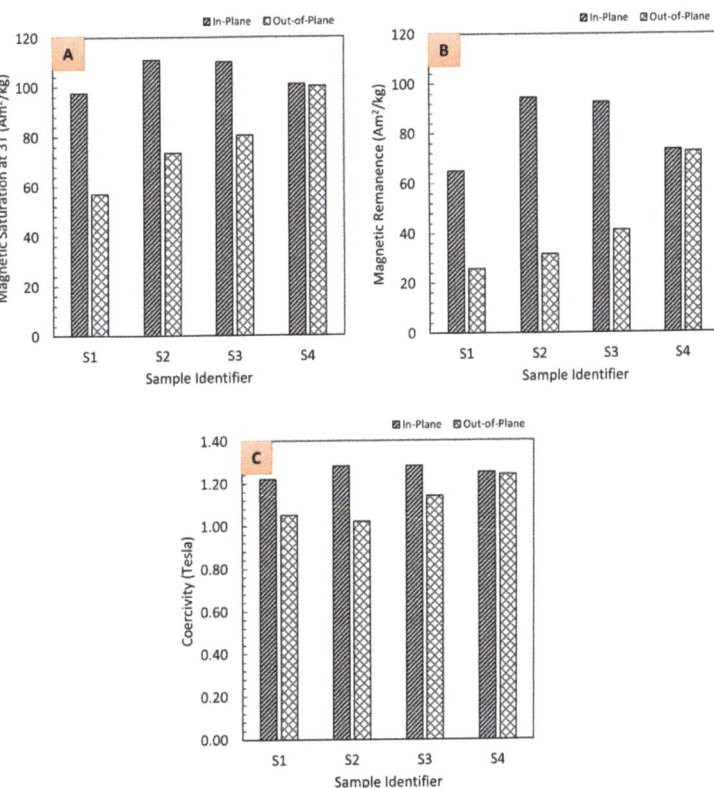

Figure 14. In-plane and out-of-plane magnetic properties of composites as listed in Table 1, derived considering magnetization and demagnetization cycles of hysteresis data: (**A**) magnetic saturation at 3 Tesla, (**B**) remanence, and (**C**) coercivity.

5. Conclusions

In this study, magnetic composites with anisotropic and isotropic properties were created using two different types of prepolymer formulations, i.e., a dual-cure (via UV irradiation and heat) acrylate-based resin and a commercially available heat curable epoxy. Both prepolymers were employed in a polymer casting process, while the dual-cure resin was also utilized in a material jetting additive manufacturing approach. It was demonstrated that the thermal initiator added to the UV curable prepolymer rendered the formulation to also be heat curable. A thermal initiator loading of 0.4 wt% was observed to solidify the prepolymer reinforced with approximately 80 wt% NdFeB magnetic particles. Molds made using stereolithography 3D printer enabled holding the resin during the magnetic alignment and thermal curing processes. The developed dual-cure prepolymer formulation was shown to be 3D printable using in-house developed material jetting AM equipment. It was observed that the thickness of cured composite layers was greater than what would be expected for the given process parameters and an unreinforced prepolymer. SEM images of magnetic composites revealed a material morphology with filler alignment and stacking features indicating directional orientation of magnetic particles in specimens fabricated using the polymer casting approach. Irrespective of the type of prepolymer used, the microstructures were highly similar. Conversely, casting the dual-cure prepolymer without the magnetic alignment process resulted in a composite sample with isotropic appearance. The microstructure of the sample fabricated using material jetting and magnetic alignment appeared to feature some particle alignment and stacking while also having similarities to the isotropic morphology. The observed morphology was attributed to a lower magnetic alignment time utilized in the additive manufacturing approach.

The ability of the permanent magnet array to orient magnetic particles in the prepolymer was confirmed through X-ray diffraction analysis where enhancements in peak intensities corresponding to the alignment of the easy axis of magnetization were observed. Thermogravimetric analysis enabled determining the magnetic filler loading in the polymer composites by conducting a thermal resin removal. From FTIR spectroscopy, through observations of peak flattening along the carbon double bond regions, it was confirmed that the UV curable resin modified with the thermal initiator was polymerized under the influence of heat.

Composites characterized for magnetic properties using a vibrating sample magnetometer revealed an enhancement in magnetic properties along the in-plane direction, i.e., the direction of magnetic field structuring, compared to the out-of-plane directions. Magnetic saturation within the tested applied magnetic field range, remanence, and coercivity were all observed to be enhanced along the in-plane direction. Samples with magnetic filler alignment and fabricated through the polymer casting approach were observed to exhibit the highest magnetic characteristics. Even though microscopy of the 3D printed sample did not reveal strong filler alignment, anisotropic magnetic properties were ascertained, albeit lower than for the samples produced by polymer casting.

This research validated the efficacy of magnetic field induced alignment of a magnetic filler for the fabrication of magnetic thermoset composites. The use of a dual-cure thermoset resin further enabled material jetting additive manufacturing using UV irradiation for pre-curing the prepolymer during printing, followed by post-processing via heat curing to ensure full polymerization of the opaque material formulation. Future work shall encompass process optimization to further enhance magnetic properties, improve dimensional accuracy, and mitigate unwanted morphological features such as resin rich zones and voids. Ultimately, this research has provided a pathway to combine a dual-cure resin formulation along with magnetic alignment and additive manufacturing to construct anisotropic magnetic composites for applications in electrical and electronic devices.

Author Contributions: B.N., A.J.Q. and P.M. conceptualized the research framework. B.N. developed the methodology and conducted the formal analysis and validation. B.N., Y.W. and M.T. performed data curation. S.T., S.B., A.J.Q. and P.M. provided facilities, supervision, funding acquisition, and project administration. B.N. prepared visualization and wrote the original manuscript draft. B.N., S.T., M.T. and P.M. performed review and editing. All authors have read and agreed to the published version of the manuscript.

Funding: This research and the APC were funded, in part, by the University of Alberta's Future Energy Systems program through the Canada First Research Excellence Fund with grant number Future Energy Systems T06-P03. In addition, B.N. received support from the Mitacs Accelerate program with award number IT16940. The authors are thankful for the gracious funding to make this research possible. M.T. received support from Postdoctoral Fellowships program of Natural Sciences and Engineering Research Council of Canada (NSERC).

Institutional Review Board Statement: Not applicable.

Informed Consent Statement: Not applicable.

Data Availability Statement: The data that support the findings of this study are available from the corresponding author upon reasonable request.

Conflicts of Interest: The authors declare no conflict of interest.

References

1. Najgebauer, M.; Szczygłowski, J.; Ślusarek, B.; Przybylski, M.; Kapłon, A.; Rolek, J. Magnetic Composites in Electric Motors. In *Analysis and Simulation of Electrical and Computer Systems*; Mazur, D., Gołębiowski, M., Korkosz, M., Eds.; Springer International Publishing: New York, NY, USA, 2018; pp. 15–28. [CrossRef]
2. Coey, J.M.D. Permanent Magnets: Plugging the Gap. *Scr. Mater.* **2012**, *67*, 524–529. [CrossRef]
3. Brown, D.; Ma, B.M.; Chen, Z. Developments in the Processing and Properties of NdFeb-Type Permanent Magnets. *J. Magn. Magn. Mater.* **2002**, *248*, 432–440. [CrossRef]
4. Ormerod, J.; Constantinides, S. Bonded Permanent Magnets: Current Status and Future Opportunities (Invited). *J. Appl. Phys.* **1997**, *81*, 4816–4820. [CrossRef]
5. Xiao, J.; Otaigbe, J.U. Polymer-Bonded Magnets: Part I. Analytic Thermogravimetry to Determine the Effect of Surface Modification on Dispersion of Nd–Fe–B Fillers. *J. Mater. Res.* **1999**, *14*, 2893–2896. [CrossRef]
6. Martin, J.; Venturini, E.; Odinek, J.; Anderson, R. Anisotropic Magnetism in Field-Structured Composites. *Phys. Rev. E* **2000**, *61*, 2818–2830. [CrossRef]
7. Volpe, V.; D'Auria, M.; Sorrentino, L.; Davino, D.; Pantani, R. Injection Molding of Magneto-Sensitive Polymer Composites. *Mater. Today Commun.* **2018**, *15*, 280–287. [CrossRef]
8. Gutfleisch, O.; Willard, M.A.; Brück, E.; Chen, C.H.; Sankar, S.G.; Liu, J.P. Magnetic Materials and Devices for the 21st Century: Stronger, Lighter, and More Energy Efficient. *Adv. Mater.* **2011**, *23*, 821–842. [CrossRef] [PubMed]
9. Li, H.; Flynn, T.J.; Nation, J.C.; Kershaw, J.; Scott Stephens, L.; Trinkle, C.A. Photopatternable NdFeB Polymer Micromagnets for Microfluidics and Microrobotics Applications. *J. Micromech. Microeng.* **2013**, *23*, 065002. [CrossRef]
10. Khosla, A.; Kassegne, S. Fabrication of NdFeB-Based Permanent Rare-Earth Micromagnets by Novel Hybrid Micromolding Process. *Microsyst. Technol.* **2015**, *21*, 2315–2320. [CrossRef]
11. Yamahata, C.; Lotto, C.; Al-Assaf, E.; Gijs, M.A.M. A PMMA Valveless Micropump Using Electromagnetic Actuation. *Microfluid. Nanofluidics* **2005**, *1*, 197–207. [CrossRef]
12. Jiang, Y.; Masaoka, S.; Uehara, M.; Fujita, T.; Higuchi, K.; Maenaka, K. Micro-Structuring of Thick NdFeB Films Using High-Power Plasma Etching for Magnetic MEMS Application. *J. Micromech. Microeng.* **2011**, *21*, 045011. [CrossRef]
13. Wang, W.; Yao, Z.; Chen, J.C.; Fang, J. Composite Elastic Magnet Films with Hard Magnetic Feature. *J. Micromech. Microeng.* **2004**, *14*, 1321–1327. [CrossRef]
14. Damean, N.; Parviz, B.A.; Lee, J.N.; Odom, T.; Whitesides, G.M. Composite Ferromagnetic Photoresist for the Fabrication of Microelectromechanical Systems. *J. Micromech. Microeng.* **2005**, *15*, 29–34. [CrossRef]
15. Yuan, Z.C.; Williams, A.J.; Shields, T.C.; Blackburn, S.; Ponton, C.B. The Production of Sr Hexaferrite Thick Films by Screen Printing. *J. Magn. Magn. Mater.* **2002**, *247*, 257–269. [CrossRef]
16. Romero, J.J.; Cuadrado, R.; Pina, E.; de Hoyos, A.; Pigazo, F.; Palomares, F.J.; Hernando, A.; Sastre, R.; Gonzalez, J.M. Anisotropic Polymer Bonded Hard-Magnetic Films for Microelectromechanical System Applications. *J. Appl. Phys.* **2006**, *99*, 08N303. [CrossRef]
17. Thévenot, J.; Oliveira, H.; Sandre, O.; Lecommandoux, S. Magnetic Responsive Polymer Composite Materials. *Chem. Soc. Rev.* **2013**, *42*, 7099. [CrossRef] [PubMed]
18. Compton, B.G.; Kemp, J.W.; Novikov, T.V.; Pack, R.C.; Nlebedim, C.I.; Duty, C.E.; Rios, O.; Paranthaman, M.P. Direct-Write 3D Printing of NdFeB Bonded Magnets. *Mater. Manuf. Process.* **2018**, *33*, 109–113. [CrossRef]
19. Yang, F.; Zhang, X.; Guo, Z.; Ye, S.; Sui, Y.; Volinsky, A.A. 3D Printing of NdFeB Bonded Magnets with SrFe12O19 Addition. *J. Alloys Compd.* **2019**, *779*, 900–907. [CrossRef]
20. Nagarajan, B.; Kamkar, M.; Schoen, M.A.W.; Sundararaj, U.; Trudel, S.; Qureshi, A.J.; Mertiny, P. Development and Characterization of Stable Polymer Formulations for Manufacturing Magnetic Composites. *J. Manuf. Mater. Process.* **2020**, *4*, 4. [CrossRef]
21. Yan, Y.; Liu, L.; Ding, C.; Nguyen, L.; Moss, J.; Mei, Y.; Lu, G.Q. Additive Manufacturing of Magnetic Components for Heterogeneous Integration. In Proceedings of the IEEE 67th Electronic Components and Technology Conference, Orlando, FL, USA, 30 May–2 June 2017; IEEE: Piscataway, NJ, USA, 2017; pp. 324–330. [CrossRef]
22. Shen, A.; Peng, X.; Bailey, C.P.; Dardona, S.; Ma, A.W.K. 3D Printing of Polymer-Bonded Magnets from Highly Concentrated, Plate-like Particle Suspensions. *Mater. Des.* **2019**, *183*, 108133. [CrossRef]
23. Shen, A.; Bailey, C.P.; Ma, A.W.K.; Dardona, S. UV-Assisted Direct Write of Polymer-Bonded Magnets. *J. Magn. Magn. Mater.* **2018**, *462*, 220–225. [CrossRef]
24. Huber, C.; Abert, C.; Bruckner, F.; Groenefeld, M.; Muthsam, O.; Schuschnigg, S.; Sirak, K.; Thanhoffer, R.; Teliban, I.; Vogler, C.; et al. 3D Print of Polymer Bonded Rare-Earth Magnets, and 3D Magnetic Field Scanning with an End-User 3D Printer. *Appl. Phys. Lett.* **2016**, *109*, 162401. [CrossRef]
25. Li, L.; Tirado, A.; Nlebedim, I.C.; Rios, O.; Post, B.; Kunc, V.; Lowden, R.R.; Lara-Curzio, E.; Fredette, R.; Ormerod, J.; et al. Big Area Additive Manufacturing of High Performance Bonded NdFeB Magnets. *Sci. Rep.* **2016**, *6*, 36212. [CrossRef] [PubMed]
26. Li, L.; Post, B.; Kunc, V.; Elliott, A.M.; Paranthaman, M.P. Additive Manufacturing of Near-Net-Shape Bonded Magnets: Prospects and Challenges. *Scr. Mater.* **2017**, *135*, 100–104. [CrossRef]
27. Parans Paranthaman, M.; Yildirim, V.; Lamichhane, T.N.; Begley, B.A.; Post, B.K.; Hassen, A.A.; Sales, B.C.; Gandha, K.; Nlebedim, I.C. Additive Manufacturing of Isotropic NdFeB PPS Bonded Permanent Magnets. *Materials* **2020**, *13*, 3319. [CrossRef]

28. Nlebedim, I.C.; Ucar, H.; Hatter, C.B.; McCallum, R.W.; McCall, S.K.; Kramer, M.J.; Paranthaman, M.P. Studies on in Situ Magnetic Alignment of Bonded Anisotropic Nd-Fe-B Alloy Powders. *J. Magn. Magn. Mater.* **2017**, *422*, 168–173. [CrossRef]
29. Gandha, K.; Li, L.; Nlebedim, I.C.; Post, B.K.; Kunc, V.; Sales, B.C.; Bell, J.; Paranthaman, M.P. Additive Manufacturing of Anisotropic Hybrid NdFeB-SmFeN Nylon Composite Bonded Magnets. *J. Magn. Magn. Mater.* **2018**, *467*, 8–13. [CrossRef]
30. Khazdozian, H.A.; Li, L.; Paranthaman, M.P.; McCall, S.K.; Kramer, M.J.; Nlebedim, I.C. Low-Field Alignment of Anisotropic Bonded Magnets for Additive Manufacturing of Permanent Magnet Motors. *JOM* **2019**, *71*, 626–632. [CrossRef]
31. Gandha, K.; Nlebedim, I.C.; Kunc, V.; Lara-Curzio, E.; Fredette, R.; Paranthaman, M.P. Additive Manufacturing of Highly Dense Anisotropic Nd–Fe–B Bonded Magnets. *Scr. Mater.* **2020**, *183*, 91–95. [CrossRef]
32. Sarkar, A.; Somashekara, M.A.; Paranthaman, M.P.; Kramer, M.; Haase, C.; Nlebedim, I.C. Functionalizing Magnet Additive Manufacturing with In-Situ Magnetic Field Source. *Addit. Manuf.* **2020**, *34*, 101289. [CrossRef]
33. Rudin, A.; Choi, P. *The Elements of Polymer Science and Engineering*, 3rd ed.; Elsevier: Boston, MA, USA, 2013; pp. 341–389.
34. Nagarajan, B.; Aguilera, A.F.E.; Qureshi, A.; Mertiny, P. Additive Manufacturing of Magnetically Loaded Polymer Composites: An Experimental Study for Process Development. In Proceedings of the ASME 2017 International Mechanical Engineering Congress and Exposition, Tampa, FL, USA, 3–9 November 2017; American Society of Mechanical Engineers: New York, NY, USA, 2017. V002T02A032.
35. Nagarajan, B.; Schoen, M.A.W.; Trudel, S.; Qureshi, A.J.; Mertiny, P. Rheology-Assisted Microstructure Control for Printing Magnetic Composites—Material and Process Development. *Polymers* **2020**, *12*, 2143. [CrossRef] [PubMed]
36. Schläfer, D.; Walker, T.; Mattern, N.; Grünberger, W.; Hinz, D. Analysis of Texture Distribution in NdFeB Hard Magnets by Means of X-ray Diffraction in Bragg-Brentano Geometry. *Texture Stress. Microstruct.* **1996**, *26*, 71–81. [CrossRef]
37. Uhl, F.M.; Webster, D.C.; Davuluri, S.P.; Wong, S.C. UV Curable Epoxy Acrylate–Clay Nanocomposites. *Eur. Polym. J.* **2006**, *42*, 2596–2605. [CrossRef]
38. Wu, K.C.; Halloran, J.W. Photopolymerization Monitoring of Ceramic Stereolithography Resins by FTIR Methods. *J. Mater. Sci.* **2005**, *40*, 71–76. [CrossRef]
39. Nagarajan, B.; Mertiny, P.; Qureshi, A.J. Magnetically Loaded Polymer Composites Using Stereolithography—Material Processing and Characterization. *Mater. Today Commun.* **2020**, *25*, 101520. [CrossRef]

Article

Mechanical Properties of PolyJet 3D-Printed Composites Inspired by Space-Filling Peano Curves

Changlang Wu [1], Truong Tho Do [2] and Phuong Tran [1,3,4,*]

[1] School of Civil and Infrastructure Engineering, RMIT University, Melbourne, VIC 3000, Australia; s3819965@student.rmit.edu.au
[2] College of Engineering and Computer Science, VinUniversity, Hanoi 14000, Vietnam; truong.dt@vinuni.edu.vn
[3] Advanced Manufacturing Precinct, School of Engineering, City Campus, RMIT University, Melbourne, VIC 3000, Australia
[4] Centre for Innovative Structures and Materials, School of Engineering, RMIT University, Melbourne, VIC 3001, Australia
* Correspondence: jonathan.tran@rmit.edu.au

Citation: Wu, C.; Do, T.T.; Tran, P. Mechanical Properties of PolyJet 3D-Printed Composites Inspired by Space-Filling Peano Curves. *Polymers* **2021**, *13*, 3516. https://doi.org/10.3390/polym13203516

Academic Editors: Swee Leong Sing and Wai Yee Yeong

Received: 31 August 2021
Accepted: 27 September 2021
Published: 13 October 2021

Publisher's Note: MDPI stays neutral with regard to jurisdictional claims in published maps and institutional affiliations.

Copyright: © 2021 by the authors. Licensee MDPI, Basel, Switzerland. This article is an open access article distributed under the terms and conditions of the Creative Commons Attribution (CC BY) license (https://creativecommons.org/licenses/by/4.0/).

Abstract: This paper proposes a design of novel composite materials inspired by the Peano curve and manufactured using PolyJet 3D printing technology with Agilus30 (flexible phase) and VeroMagentaV (rigid phase) materials. Mechanical properties were evaluated through tensile and compression tests. The general rule of mixture (ROM) for composites was employed to approximate the tensile properties of the hybrid materials and compare them to the experimental results. The effect of reinforcement alignments and different hierarchies are discussed. The results indicated that the 5% inclusion of the Peano reinforcement in tensile samples contributed to the improvement in the elastic modulus by up to 6 MPa, but provided no obvious enhancement in ultimate tensile strength. Additionally, compressive strengths between 2 MPa and 6 MPa were observed for compression cubes with first-order reinforcement, while lower values around 2 MPa were found for samples with second-order reinforcement. That is to say, the first-order reinforcement has been demonstrated more effectively than the second-order reinforcement, given the same reinforcement volume fraction of 10% in compression cubes. Different second-order designs exhibited slightly different mechanical properties based on the ratio of reinforcement parallel to the loading direction.

Keywords: Peano curve; composite; PolyJet 3D printing; rule of mixture; multi-material printing; additive manufacturing

1. Introduction

Fractal patterns exist everywhere in nature in various ways, such as in spider webs, the Milky Way galaxy, and coastlines. The concept of fractal was first introduced by Mandelbrot [1] in 1977. He defines it in the book *Fractals in Physics* as [2]:

'Fractal is a structure comprised of parts that, in some manner, are similar to the whole of this structure.' (p. 250)

Self-similarity, the main attribute of fractal patterns, indicates that the geometry consists of a unit structure repeating itself in different scales [3]. The self-similarity feature can be found in many objects, such as Russian matryoshka dolls, the Koch snowflake, etc. However, fractal structures were not applied to industries until some theoretical analyses and experiments were conducted recently [4]. Space-filling curves are special cases of fractal structures, which are characterized by a unique property that, after an infinite number of iterations, a finite area would be filled with a curve of infinite length. The most famous space-filling curves include the Peano curve, the Hilbert curve, and the Moore curve.

In the past two decades, scientists have embraced the study of fractal geometries, with respect to electronics design (Figure 1a). Studies reveal that fractal-shaped antennas show superior properties from their geometrical attributes. The self-similarity characteristic of fractal patterns contributes to a multiband feature of the corresponding antennas [5–8], while the high convoluted shape and space-filling properties of certain fractal curves allow for the reduction of the miniaturization of microstrip antennas, resonators, and filters [9–12]. These properties show great potential for designing multiband antennas, frequency-selective surfaces, and reducing the size of antennas. Since fractal geometry was first introduced to antenna array design by Kim and Jaggard [13], various space-filling curve designs have been utilised to improve the performance of antennas, including the Peano curve [14–18], Hilbert curves [14–16,19], the Koch curve [8,20,21], the Gosper curve [22,23], the Moore curve [10,16], the Sierpinski curve [6,7], the Minkowski curve [24,25], the Greek cross [16], and combinations of multiple geometries, such as the Peano-Gosper curve [26–28], the Koch-Sierpinski shape [9,29,30], and the Hilbert-Minkowski pattern [31]. Moreover, the mechanical stretchability of space-filling shaped electronics has attracted growing interest from researchers to achieve both advanced electronic function and compliant mechanics. Fan et al. [16] demonstrated that fractal-based structures bonded to pre-strained elastomers enable higher levels of elastic deformation. It was also indicated that fractal-based layout could provide a strategy to integrate hard and soft materials. Similar studies were conducted to investigate the stretchability of fractal-based stretchable electronics [32–35].

In addition to its value in electronics, fractal-based geometry has also been adapted for novel material design in recent studies. Fractal patterns appear in many natural materials, such as shells and bones. These natural materials have attracted considerable attention from scientists due to their excellent mechanical properties. Huiskes et al. [36] claimed that the fractal morphology of trabecular bone contributed partly to its mechanical efficiency. Following this theory, Farr [37–39] applied fractal principles to structure designs, showing the improvement in mechanical efficiency under gentle compressive loading conditions. So far, many studies have been conducted on fractal-like hierarchical honeycombs regarding both in-plane and out-of-plane properties [40–49]. In 2015, Meza et al. [50] created structural metamaterials with exceptional strength, stiffness, and damage tolerance from materials in which unit cells were organized into a self-repeating geometry. Wang et al. [51] proposed a Koch-curve hybrid structure as shown in Figure 1b, indicating its energy absorption capability and lightweight feature. Additionally, fractal-like patterns have also been demonstrated to be promising in the design of stronger interlockings. Typical examples are the hierarchical suture joints inspired by ammonite fossils [52] and the 3D-printed Koch curve interlockings [53]. It was shown that the load-bearing capacity of the interlocking could be effectively increased via fractal design. Recently, the well-known 3D fractal structures, which are called Menger Sponge cubes, were 3D printed using direct laser lithography [54] and demonstrated superior energy absorption ability.

The emergence of additive manufacturing realizes the fabrication of structures with complex geometries and exceptional engineering properties, which could not be achieved by conventional manufacturing methods. Recent studies regarding multi-material 3D printing have demonstrated its superior function in creating structures/materials with tunable mechanical properties [55]. For example, multi-material fused deposition modelling (mFDM) 3D printing technology was utilised by Zhang et al. [56] to manufacture functionally gradient composites with user-defined mechanical properties. More studies have been conducted using material-jetting technology. In 2019, Skylar-Scott et al. [57] proposed an inkjet multi-material, multi-nozzle 3D printing method to generate origami structures, using two different viscoelastic epoxy inks for flexible hinges and rigid faces, respectively. The resulting structures showed the capability during compression in terms of large deformation in the hinges and multiple folding cycles before failure. Later, Yuan et al. [58] used PolyJet technology to fabricate composites with two photopolymers, VeroBlack and TangoPlus. According to their study, programmed shape-memory behaviours were achieved by the 3D printed structures.

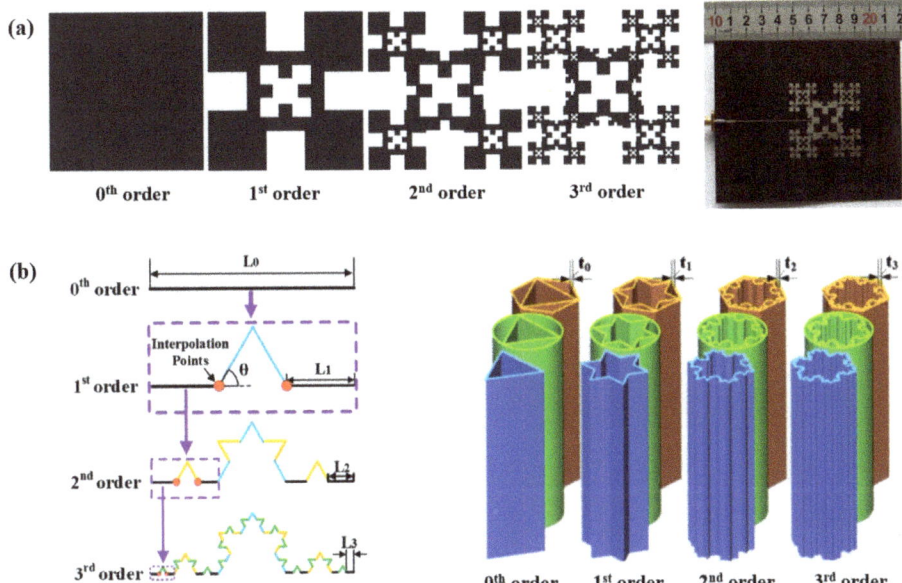

Figure 1. (a) Koch and Sierpinski patterns inspired antenna design [9], reproduced courtesy of The Electromagnetics Academy; (b) Koch snowflake inspired thin-walled structure design for energy absorption [51].

Previous studies have successfully demonstrated the potential of fractal patterns in material design, whereas a limited variety of self-similar shapes have been explored. Despite the fact that multi-material printing exhibits the capability to create structures/materials with tunable properties, most studies focused on single material design and fabrication. In this study, we propose a novel design of 3D-printed composites. The hybrid materials feature a space-filling curve modified reinforcement and are manufactured using PolyJet 3D printing technology. Experiments, microscopy, and analytical models are conducted to investigate the mechanical properties of innovative materials. The results of this study provide insight into a continuous-curve-reinforced polymer composite, which has potential application in biomedical [59], automotive [60], and aerospace engineering [61].

2. Methods

2.1. Material Design and Fabrication

The Peano curve, which was introduced by an Italian mathematician Giuseppe Peano, was the first space-filling curve to be discovered. The set of curves consists of many orders, which can be constructed following a sequence of steps as shown in Figure 2a. Considering mechanical properties and manufacturing issues, all the sharp edges in the original Peano curves are smoothed using arcs as shown in Figure 2b. Rhino with the Grasshopper plugin is employed as the CAD software. Figure 2b defines three geometric parameters, i.e., the side length of a small square (l), arc curvature (k), and diameter (D). Six patterns of Peano curves are to be investigated in this study with respect to various orientations and different hierarchies as shown in Figure 2c.

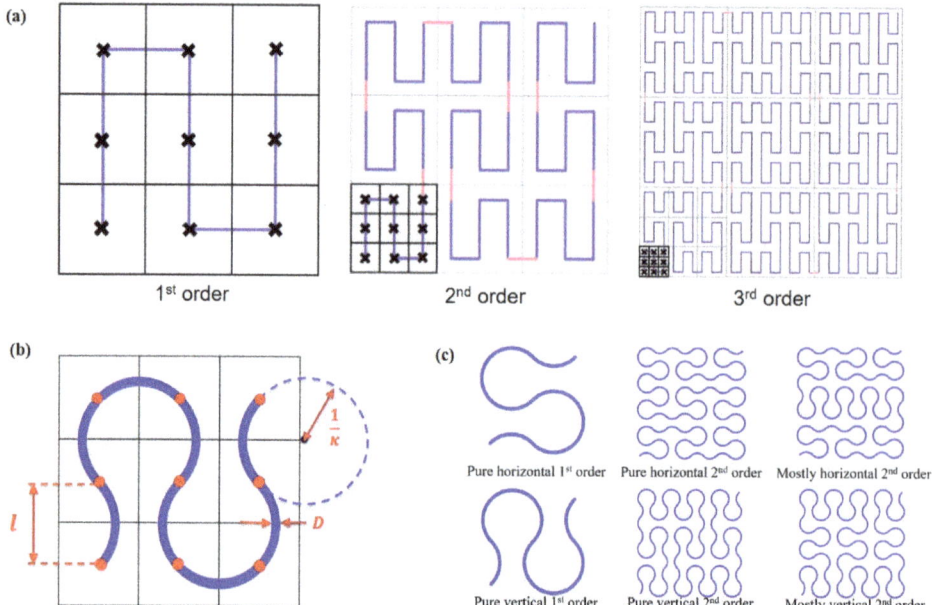

Figure 2. (a) Construction of the first three orders of Peano curves; (b) schematic design of a smoothed Peano curve, including control points (red dots), the side length of a small square (l), arc curvature (k), and diameter (D); (c) variants of Peano curves at different orders to be investigated in this study.

The proposed first order and second order Peano curves are designed to act as a hard reinforcement, which is embedded in a soft-material matrix in order to investigate the mechanical performance of 3D-printed composites. Stratasys J750 Digital Anatomy 3D printer, provided by Stratasys Ltd., Rehovot, Israel, is a PolyJet 3D printer and was used to fabricate all the samples. This printer has four inkjet heads and two UV light sources, allowing multi-material 3D printing from a wide range of available materials. J750 is also capable of generating complex geometries with microscopic layer resolution, down to 0.014 mm. All the samples were manufactured with two different materials, VeroMagentaV (VMV) and Agilus30 (A30). VMV is a rigid and opaque photopolymer, while A30 is a rubber-like polymer. VMV is from the family of Vero; available in seven hues, including blue, white, black, grey, cyan, magenta, and yellow, the Vero family shares similar mechanical, thermal, and electrical properties. Here, VeroMagentaV is selected to offer a more saturated and vibrant colour compared to the transparent A30.

So far, no standard of the tensile test has been established for 3D-printed multi-material structures/materials. In this study, tensile samples are designed according to the ASTM D638, with variations from the literature [62,63] as shown in Figure 3a. Two categories of samples are prepared for further analysis. First, homogeneous A30 samples are printed to capture their individual mechanical properties, thereby providing a reference to composite materials. Then, six designs of hybrid samples (Figure 3b) are fabricated, with A30 serving as the matrix of gauge section, and VMV as both the reinforcements and extended sections. Figure 3b schematically shows the gauge sections of six heterogenous designs, reinforced with differently orientated and hierarchical Peano curves, including pure vertical first order, pure horizontal first order, pure vertical second order, pure horizontal second order, mostly vertical second order, and mostly horizontal second order. The reinforcements are distributed in three layers at a spacing of 1 mm. All the hybrid tensile structures were reinforced with VMV at a volume fraction of 5%. Thus, the diameters for the first and second order Peano reinforcements are 0.36 mm and 0.212 mm, respectively. Figure 3c depicts the 3D-printed tensile samples.

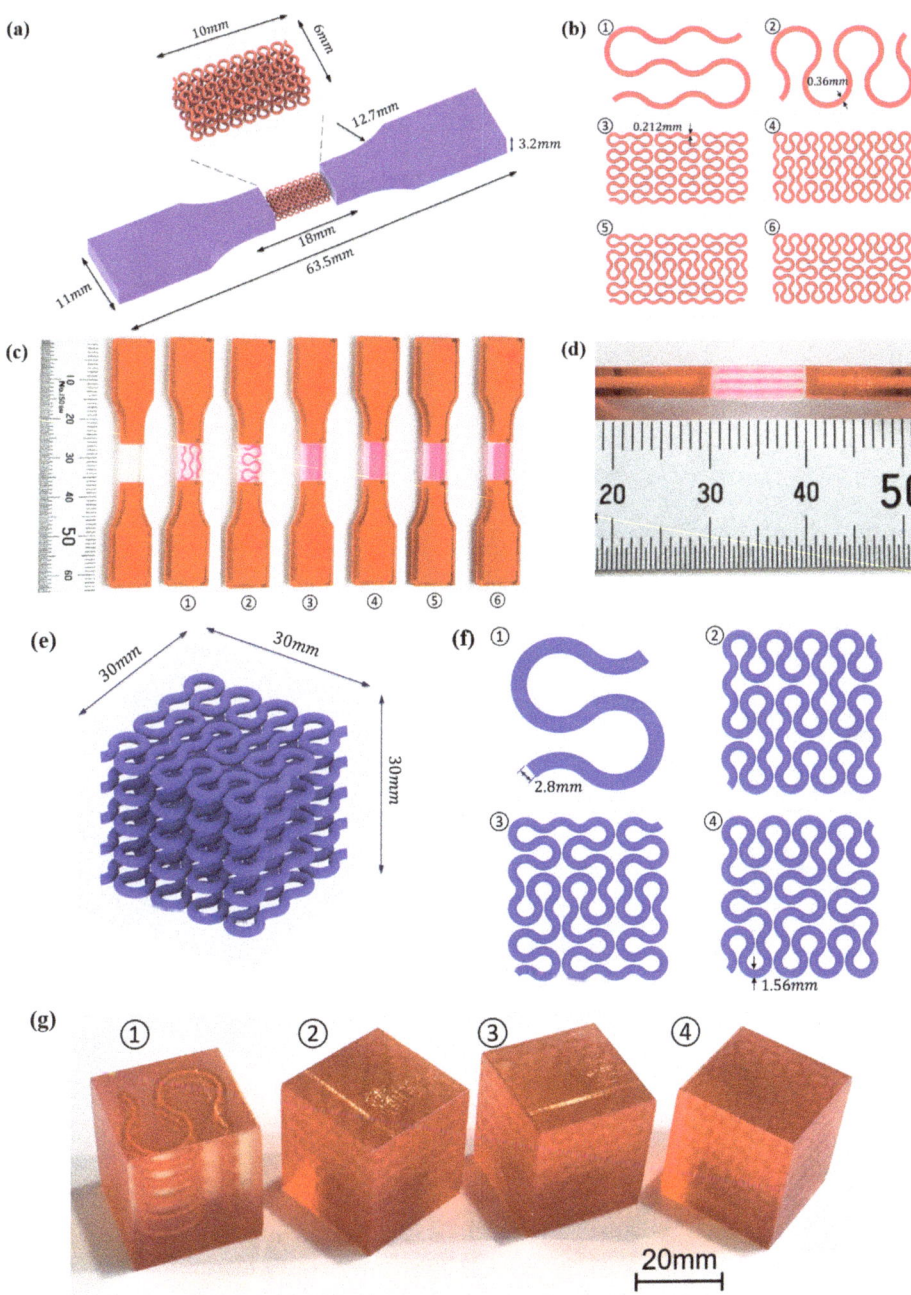

Figure 3. (a) Dimensions of a tensile specimen; (b) plan views of six different hybrid structures using different Peano patterns; (c) 3D-printed tensile samples, including pure A30 and composites with VMV reinforcement embedded inside A30; (d) side view of the gauge section for the hybrid materials, showing the three-layer reinforcement; (e) dimensions of the compression

cube; (**f**) plan views of four different hybrid structures showing the first order reinforcement with a reinforcement diameter of 2.8 mm and 1.156 mm for all three second order designs; (**g**) pictures of 3D printed samples (from left to right, corresponds to case 1 to case 4, respectively).

The compression specimens are designed as cubes with a side length of 30 mm, as shown in Figure 3e. The cubic matrix is A30, which is reinforced by five-layer VMV Peano curves at a spacing of 5 mm. Four different infills, with a volume fraction of 10%, are introduced as shown in Figure 3f,g. The first order Peano reinforcement has a diameter of 2.8 mm, while the second order has a diameter of 1.56 mm.

2.2. Mechanical Testings

In order to investigate the mechanical properties of Peano reinforced hybrid materials, tensile and compression tests were conducted using the Universal Instron testing machine. Tensile tests were controlled with a displacement rate of 1 mm/min until a failure happens, while the uniaxial compression tests were performed with a rate of 1.3 mm/min until strain reaches 60%. Compression tests were performed from three axial directions (Table 1) considering the anisotropic property of the cubic designs. Fives samples for each type of design were tested to minimise the experimental artifacts.

Table 1. Schematic diagrams and experimental pictures showing three different compressive loading directions (taking the case 2 design as a schematic example).

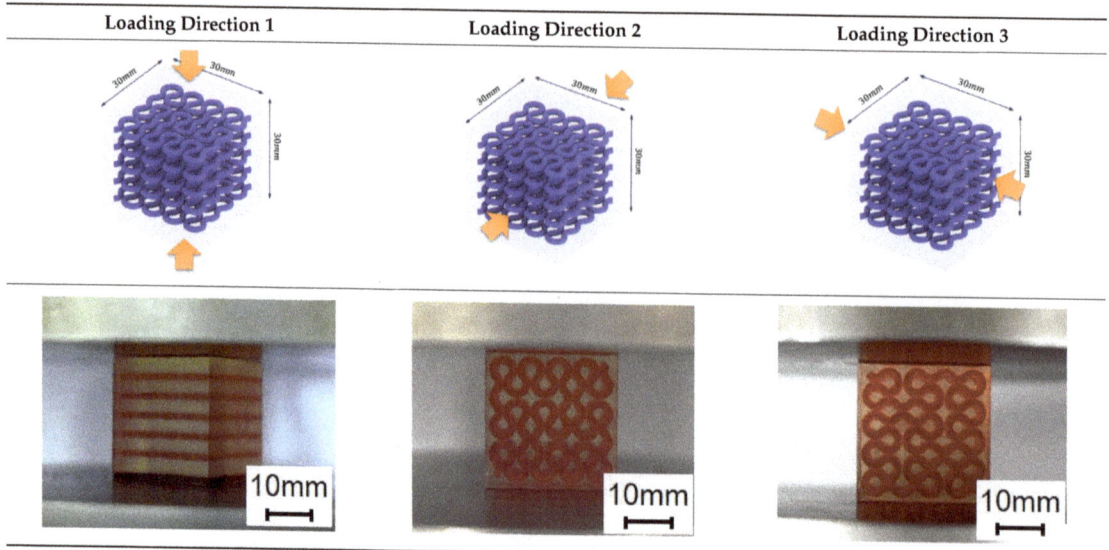

2.3. Rule of Mixture for Composites

In order to provide theoretical references for experimental results, the Rule of Mixture (ROM) was adapted in this study to approximate the elastic properties of composite materials. Based on different assumptions, both the upper and lower bounds of the elastic modulus for composites could be found. When the load is applied longitudinally to the fibre, the ROM defines the highest elastic modulus of the composite according to the iso-strain assumption:

$$E_{c,\ max} = fE_f + (1-f)E_m, \qquad (1)$$

where $E_{c,min}$ denotes the upper bound of the elastic modulus of the composite; $f = \frac{V_f}{V_f + V_m}$ is the volume fraction of reinforcement; E_f is the elastic modulus of the VMV reinforcement;

and E_m is the elastic modulus of the A30 matrix. It should be noted that Equation (1) can also be applied to predict other elastic properties, for example, the ultimate tensile strength.

When the load is applied transverse to the fibre, the lower bound of the elastic modulus could be estimated using the following equation according to the iso-stress assumption:

$$E_{c,min} = \left(\frac{f}{E_f} + \frac{1-f}{E_m}\right)^{-1}, \quad (2)$$

where $E_{c,min}$ denotes the lower bound of the elastic modulus of the composite.

In this study, the theoretical range of the elastic moduli of the novel hybrid materials is predicted by Equation (1) and Equation (2). The experimental results are expected to sit in between the range. Additionally, the ultimate tensile strength of the composite materials is estimated using the ROM by Equation (1). In the next section, the approximations from the analytical models and experiments are compared with detailed discussions on the discrepancy.

3. Results and Discussion

3.1. Tensile Test Results and Discussion

Herein, stress-strain curves obtained in tensile tests are presented and compared. Five specimens for each design were tested and the results are illustrated with details of the average stress and standard deviation. The elastic moduli and ultimate tensile strengths are captured from the experiments and then compared to theoretical estimations. Moreover, crack propagations and fracture surfaces are investigated with representative microscope images provided.

Figure 4 shows the tensile testing results for pure A30 samples. Despite the slightly different elongations of the five specimens, the non-linear responses of all five tests are repeatable. As revealed by Figure 4a, specimen two experiences the maximum stress of 0.94 MPa, while specimen four experiences the least at 0.85 MPa.

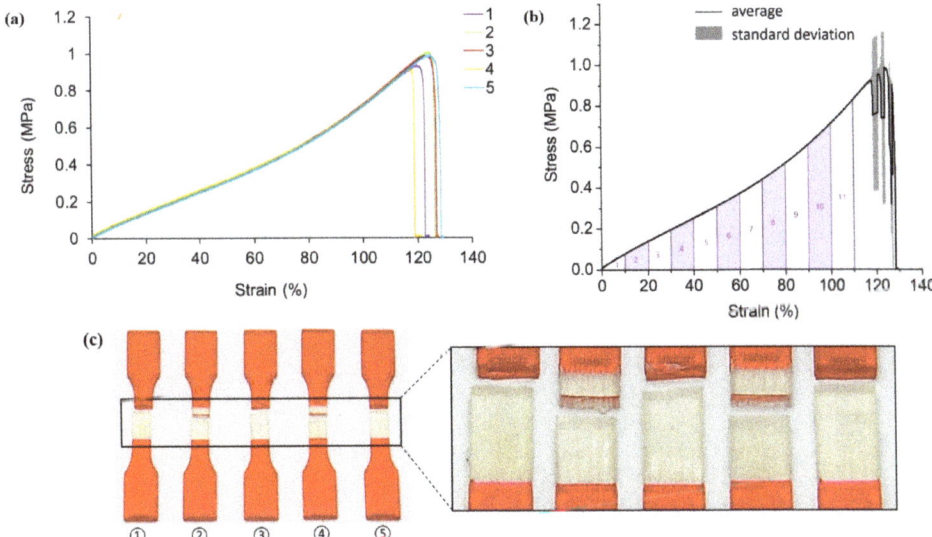

Figure 4. (a) Stress-strain curves obtained from tensile testing tests on non-reinforced A30 samples; (b) average tensile stress—strain curve and standard deviation of all five testings; (c) pictures of failed tensile samples with zoom-in at the locations of fracture.

Since A30 is a rubber-like, hyper-elastic material, it is typically not described using Young's modulus and Poisson's ratio [64]. To be more specific, the elastic modulus of A30 is not constant but changes with strain. In order to approximate the value, the average stress-strain curve before fracture is divided into eleven segments. Each segment corresponds to a 10% strain change as shown and numbered in Figure 4b. The stress-strain curve within each segment is assumed to be linear so that the elastic moduli of A30 could be estimated. Results from the eleven segments approximate a range from 0.56 MPa to 1.18 MPa for the elastic modulus of A30.

Figure 4c depicts the failure samples with a magnified picture at the gauge sections. Fractures are identified to happen at different locations, including the gauge section (specimen two and four from left to right), close to the extension part (specimen one and three from left to right), and also at the interface of two different materials (specimen five). This phenomenon could ascribe to 3D printing defects.

Figure 5 shows the tensile test results of hybrid case one samples, which introduce the first order pure horizontal Peano VMV reinforcement into the A30 matrix. The responses of all five specimens are similar, particularly the elastic deformation stage (strain less than 20%) as suggested by the stress-strain curves in Figure 5a. All specimens experience similar maximum tensile stress of 1 MPa, approximately.

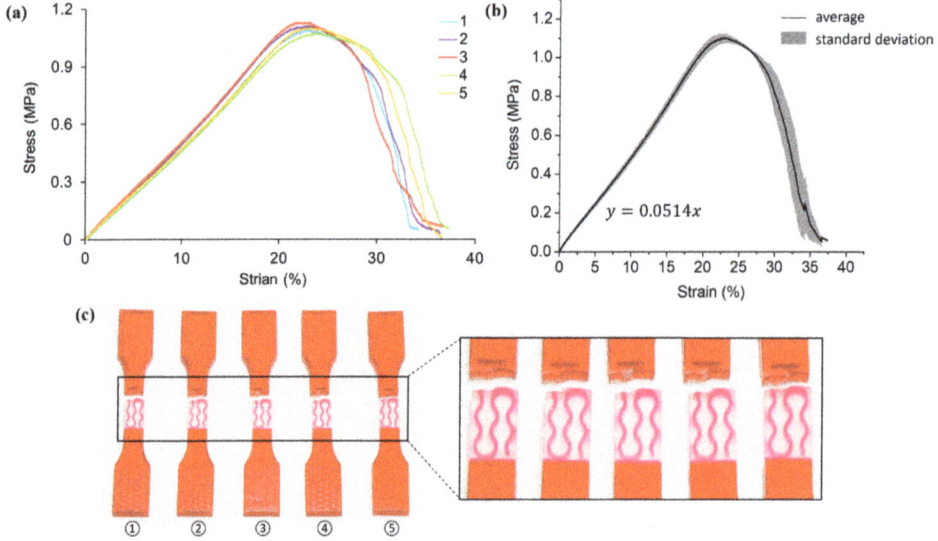

Figure 5. (a) Stress-strain curves obtained from tensile testing tests on hybrid case one (pure horizontal first order) samples; (b) average tensile stress-strain curve and standard deviation of all five testings; (c) pictures of failed tensile samples with zoom-in at the locations of fracture.

An average elastic modulus of 5.14 MPa is captured in Figure 5b. Compared to the results of homogenous A30 samples, there are improvements in both the ultimate tensile stress and elastic modulus. As the results imply, the introduction of embedded VMV reinforcement in hybrid case one contributes to an enhancement in both the tensile strength and stiffness.

Different from homogenous A30 samples, all the fractures of case one samples are located in the A30 matrix and near the edge of the gauge section (Figure 5c). In other words, failure only happens between the edge of the reinforcement and the extension. This phenomenon could be explained by the non-effective stress transfer between A30 and VMV. According to the material datasheet provided by Stratasys Ltd., VMV has a much higher strength and stiffness than A30. As a result, crack would initiate in A30 instead of VMV after the elastic deformation phase. Given the fact that there is no reinforcement existing near the extensions, these cross-sections are the most vulnerable when subjected to tensile force. Therefore, the crack initiates and propagates in the matrix near the extension until it totally fails.

Figure 6 describes the tensile test results on the hybrid case two samples, featuring the first order pure vertical Peano curve.

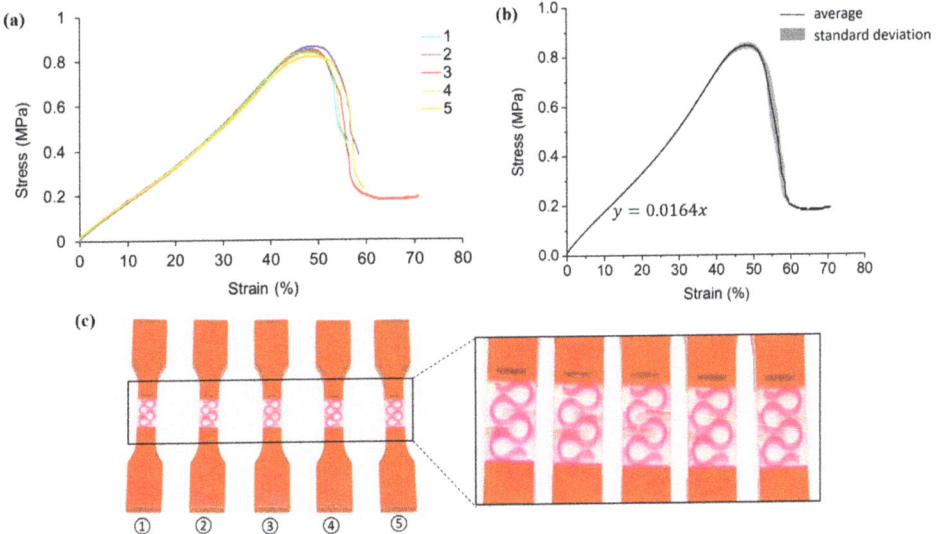

Figure 6. (a) Stress-strain curves obtained from tensile testing tests on hybrid case two (pure vertical first order) samples; (b) average tensile stress-strain curve and standard deviation of all five testings; (c) pictures of failed tensile samples with magnification at the locations of fracture.

Similar to hybrid case one, the elastic responses of all five specimens are consistent (less than 40% strain) as indicated in Figure 6a. An average elastic modulus of 1.64 MPa is identified in Figure 6b. The elastic modulus of hybrid case two is increased by 0.96 MPa compared to pure A30, which is attributed to the introduction of reinforcement. However, hybrid case two is less stiff than case one. Given that both case one and case two have the same hierarchy and volume fraction of reinforcement, it could be inferred that the orientation of the Peano curves has a significant influence on the elastic modulus.

As the gaps between the curved reinforcement are bigger than those between the reinforcement and extensions, the A30 within the reinforcement gaps is more vulnerable. Therefore, fractures of the second case happen in A30 in between the curved reinforcements (Figure 6c) rather than near the extensions as in case one. With respect to maximum tensile stress, all specimens experience similar values of around 0.8 MPa. Unlike the hybrid case one design, the strength of hybrid case two is lower than pure A30 samples. In hybrid materials, crack initiates in A30 in between the reinforcement and propagates until getting close to the reinforcement. In homogenous A30 samples, crack keeps propagating until a fracture happens since there is no reinforcement at any cross-section. However, the VMV reinforcement along the loading direction in hybrid case 2 confines the deformation of A30 in the transverse direction. Therefore, the ultimate tensile strength decreases compared to the homogeneous A30 samples.

The tensile test results of hybrid case 3 samples are exhibited in Figure 7. The stress-strain curves (Figure 7a) reveal that all specimens experience the same stress roughly before the strain reaches 10%. The average elastic modulus is captured to be 7.21 MPa as shown in Figure 7b. Similar to hybrid case one and case two, the inclusion of VMV reinforcement in case three improves the structural stiffness of the coupon samples. As indicated by the comparison between cases one and three, the second order reinforcement contributes more to the stiffness than the first order reinforcement.

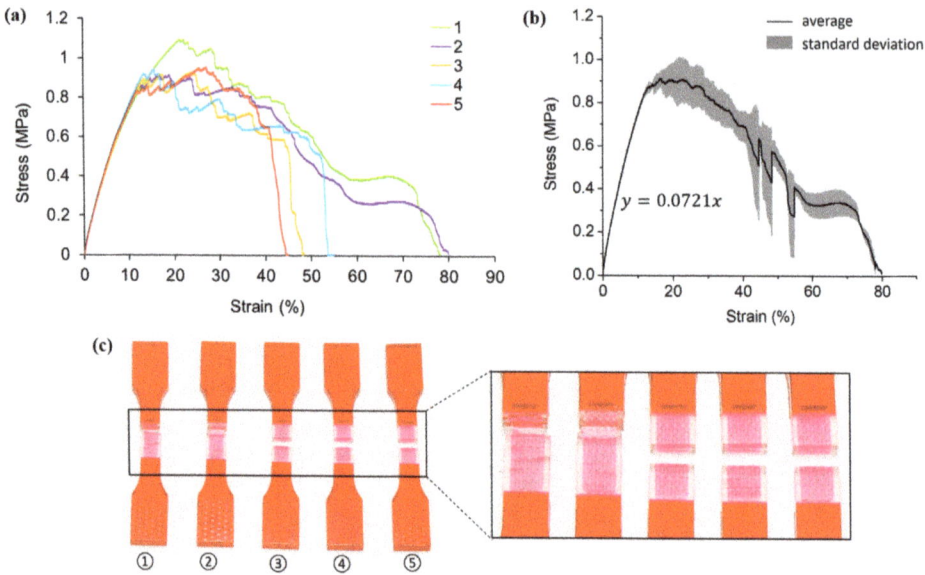

Figure 7. (**a**) Stress-strain curves obtained from tensile testing tests on hybrid case three (pure horizontal second order) samples; (**b**) average tensile stress-strain curve and standard deviation of all five testings; (**c**) pictures of failed tensile samples with zoom-in at the locations of fracture.

In the plastic deformation stage, the responses of different specimens are significantly different as described by the large standard deviation in Figure 7b. Specimen one experiences the largest maximum stress of 1.03MPa, while the others share an average around 0.88MPa. It seems that the inclusion of second-order pure horizontal reinforcement contributes little to the ultimate tensile strength compared to homogenous A30. Additionally, specimens one and two reach much greater elongations than the rest. It is worth noticing c that specimen one and two break near the extensions, while the other three fail closer to the middle of the gauge sections. Different fracture locations are mainly owed to the manufacturing defects.

Different from smooth stress-strain curves obtained for pure A30, hybrid case one, and case two, the strongly jagged pattern of stress-strain curves is observed for all case three specimens in the plastic deformation phase. Once the crack initiates in A30, it propagates perpendicular to the tensile force direction until it encounters the VMV reinforcement. Due to the arc design of the Peano curve, the straightening of the reinforcement is involved first and followed by material stretching. This process leads to a decrease in stress and a certain amount of increase afterward. Since the reinforcement design in case three is more complicated than that of case one and case two, cracks happen and develop at more cross-sections, round after round. Consequently, the stress-strain curves, after the crack initiation, are wavy until fractures happen.

Similarly, the stress-strain curves (Figure 8a) for the elastic stage are repeatable for hybrid case four samples. An average elastic modulus of 5.62 MPa is captured (Figure 8b). Comparing with the homogeneous A30 (0.56 ~1.18 MPa), the design of VMV reinforcement in case four significantly improves the structural stiffness. Again, the higher elastic modulus captured for case four than case two demonstrates that the second order reinforcement contributes more to the stiffness than the first order designs.

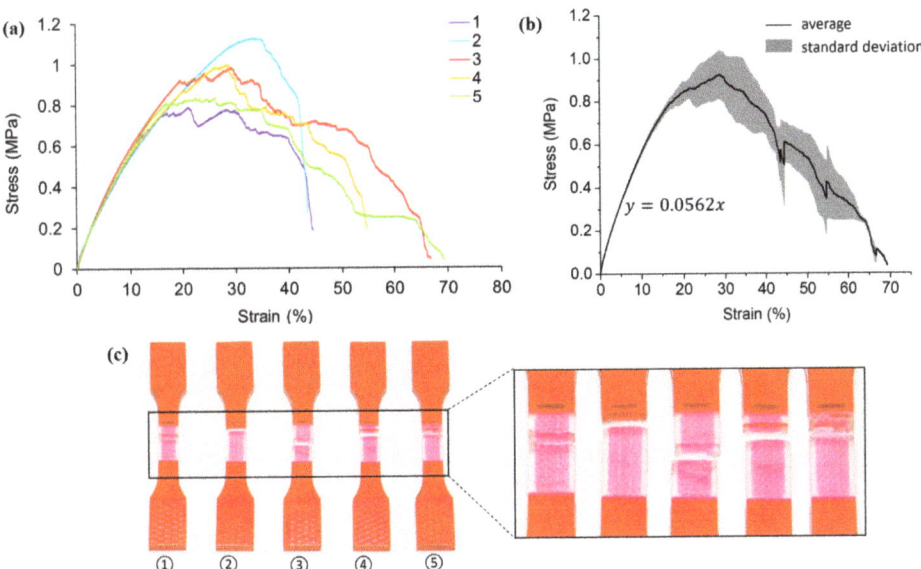

Figure 8. (a) Stress-strain curves obtained from tensile testing tests on hybrid case four (pure vertical second order) samples; (b) average tensile stress-strain curve and standard deviation of all five testings; (c) pictures of failed tensile samples with zoom-in at the locations of fracture.

The five specimens experience different ultimate tensile stress, ranging from around 0.75 MPa to 1.1 MPa. Specimen two elongates the least (45%) with the maximum ultimate tensile stress. As observed from Figure 8c, specimen two fails instead of within the gauge section but nearly at the interface of the gauge section and extension. That is to say, the entire gauge section deforms elastically until crack happens at the A30 cross-section close to the extension. Moreover, the elastic deformation stage of specimen two ends at the strain of 30%, which is longer than the other four specimens. The plastic response of specimen two is dominant to A30, thereby resulting in a less jagged stress-strain curve compared to the others. Differently, cracks happen earlier and develop at the cross-section with VMV reinforcement for specimens one, three, four, and five. This explains the lower ultimate tensile stress and wavy stress-strain curves experienced by these four specimens. As the results demonstrate, the second order pure vertical reinforcement design contributes little to the ultimate tensile strength.

Figure 9 depicts the tensile test results of the hybrid case five samples, which includes the second order mostly horizontal Peano VMV reinforcement into the A30 matrix. The repeatable response before a strain of 10% (Figure 9a) captures an average elastic modulus of 7.23 MPa (Figure 9b). Comparing to pure A30 (0.56~1.18 MPa), the design of VMV reinforcement in case five significantly improves the structural stiffness of the coupon samples.

However, the ultimate tensile stress experienced by all five specimens is not obviously increased in hybrid case five. As illustrated in Figure 9c, all specimens fail within the gauge section and at the cross-section with VMV reinforcement. Nevertheless, the elongation of different specimens varies widely between a strain of 40% to 90%. It could be observed from the magnified picture of specimen five that no obvious crack happens at other cross-sections except for the final failure. As a result, specimen five fractures at the smallest strain. On the other hand, specimen three experiences the greatest elongation of 90% with many cracks at different cross-sections.

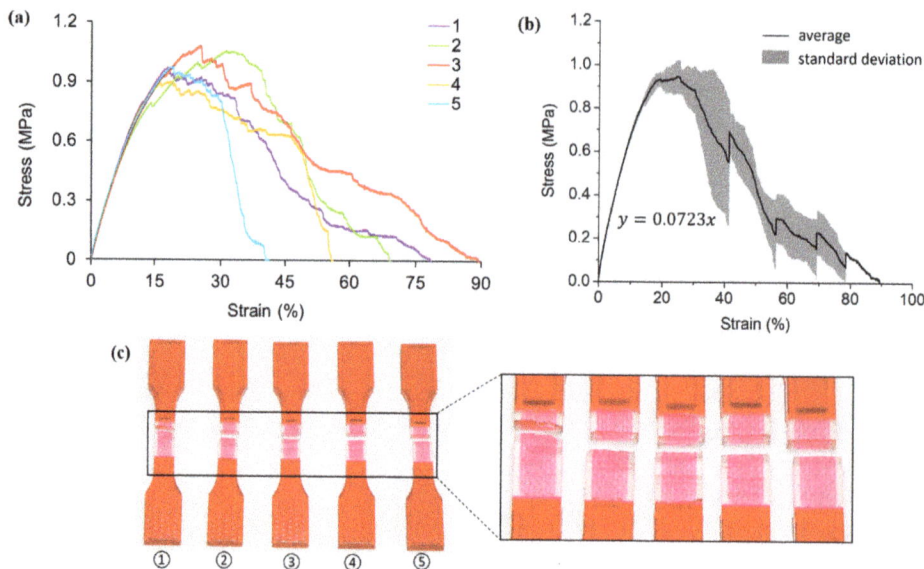

Figure 9. (**a**) Stress-strain curves obtained from tensile testing tests on hybrid case five (mostly horizontal second order) samples; (**b**) average tensile stress-strain curve and standard deviation of all five testings; (**c**) pictures of failed tensile samples with magnification at the locations of fracture.

The tensile stress-strain curves and failed samples of the hybrid case six design are presented in Figure 10. The elastic stage, corresponding to a strain less than 10%, is quite repeatable for all specimens (Figure 10a) with an average elastic modulus of 6.95 MPa (Figure 10b). Obviously, the VMV reinforcement in case six enhances the stiffness of the coupon samples compared to homogenous A30. The elastic moduli of case four and case six are very close, owing to similar reinforcement alignments (orientation and hierarchy).

The five specimens experience different ultimate tensile stress as well as elongations. Specimen one reaches the highest fracture strain of more than 90%, while its ultimate stress is the smallest among all at 0.82 MPa. On the contrary, specimen four exhibits the highest strength at 1.01 MPa and goes through the least elongation. Furthermore, specimen four fails near the extension whereas others fail within the gauge section at the cross-section containing reinforcement (Figure 10c). The wavy patterns of all stress-strain curves could be explained by the crack propagation from A30 to VMV as mentioned before. Different fracture locations are likely to result from manufacturing defects within the gauge section.

Herein, Table 2 compares the final experimental results with the analytical predictions on elastic modulus and ultimate tensile strength of the hybrid materials. Figure 11a schematically summaries the responses of materials subjected to tensile loadings. Data regarding VMV is adopted from Tee et al. [62] to help better understand the mechanical properties of the novel hybrid materials.

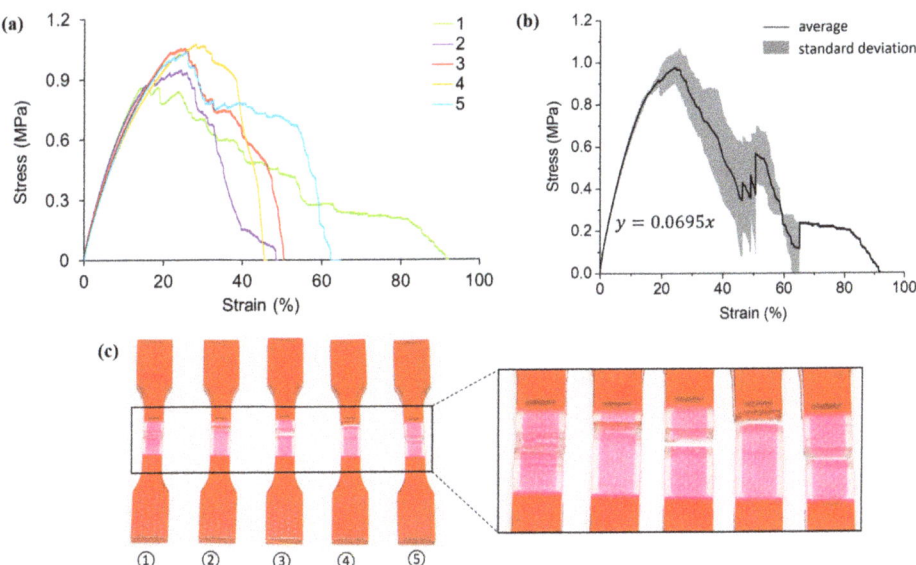

Figure 10. (a) Stress-strain curves obtained from tensile testing tests on hybrid case six (mostly vertical second order) samples; (b) average tensile stress-strain curve and standard deviation of all five testings; (c) pictures of failed tensile samples with zoom-in at the locations of fracture.

As revealed by the table, elastic moduli obtained from experiments are within the analytical prediction ranges but much closer to the lower bounds. All the hybrid materials exhibit higher elastic moduli than homogeneous A30, indicating the positive effect of VMV reinforcement embedded in the A30 matrix. For the hybrid materials with the same hierarchical reinforcements, the higher ratio of the reinforcement parallel to the loading direction and perpendicular to the loading direction leads to a higher elastic modulus. However, it is not applicable to materials with different order reinforcements. Samples reinforced by the first order Peano curves (case one and case two) yield smaller elastic moduli than the second ones (case 3–6), even though the ratio for the former ones is higher than the latter. Results demonstrate that the second order reinforcement designs are more effective than the first order despite having the same volume fraction (5%). In addition, hybrid case three and case five exhibit the highest stiffness among all. It can be concluded that the second order pure horizontal and the second order mostly horizontal reinforcement are the most effective designs in terms of stiffness enhancement.

Table 2. Comparisons between experimental results and predictions using the ROM of average elastic modulus and ultimate tensile strength.

Material		Homogeneous		Composites					
		A30	VMV	Case 1	Case 2	Case 3	Case 4	Case 5	Case 6
Ratio of reinforcements parallel and perpendicular to the loading direction		-	-	6.00	0.40	2.35	0.34	1.21	0.68
Approximate volume fraction of reinforcement parallel to the loading direction		-	-	4.28%	1.43%	3.51%	1.27%	2.73%	2.02%
E (MPa)	Experiment	0.56~1.18	858	5.14	1.64	7.21	5.62	7.23	6.95
	ROM Voigt's (upper bound)	-	-	37.31~37.91	12.81~13.42	30.65~31.25	11.47~12.09	23.99~24.60	17.87~18.48
	ROM Reuss' (lower bound)	-	-	0.59~1.22	0.59~1.20	0.58~1.22	0.57~1.19	0.57~1.21	0.57~1.20
UTS (MPa)	Experiment	0.90	57.50	1.0340	0.79	0.91	0.89	0.93	0.94
	ROM	-	-	3.32	1.71	2.89	1.62	2.45	2.04
	Discrepancy	-	-	68.86%	53.57%	68.65%	45.31%	61.88%	54.12%

Note: The ROM denotes the general rule of mixture for composites; E denotes the elastic modulus measured from tensile testing; UTS denotes the average ultimate tensile strength measured from tensile testing.

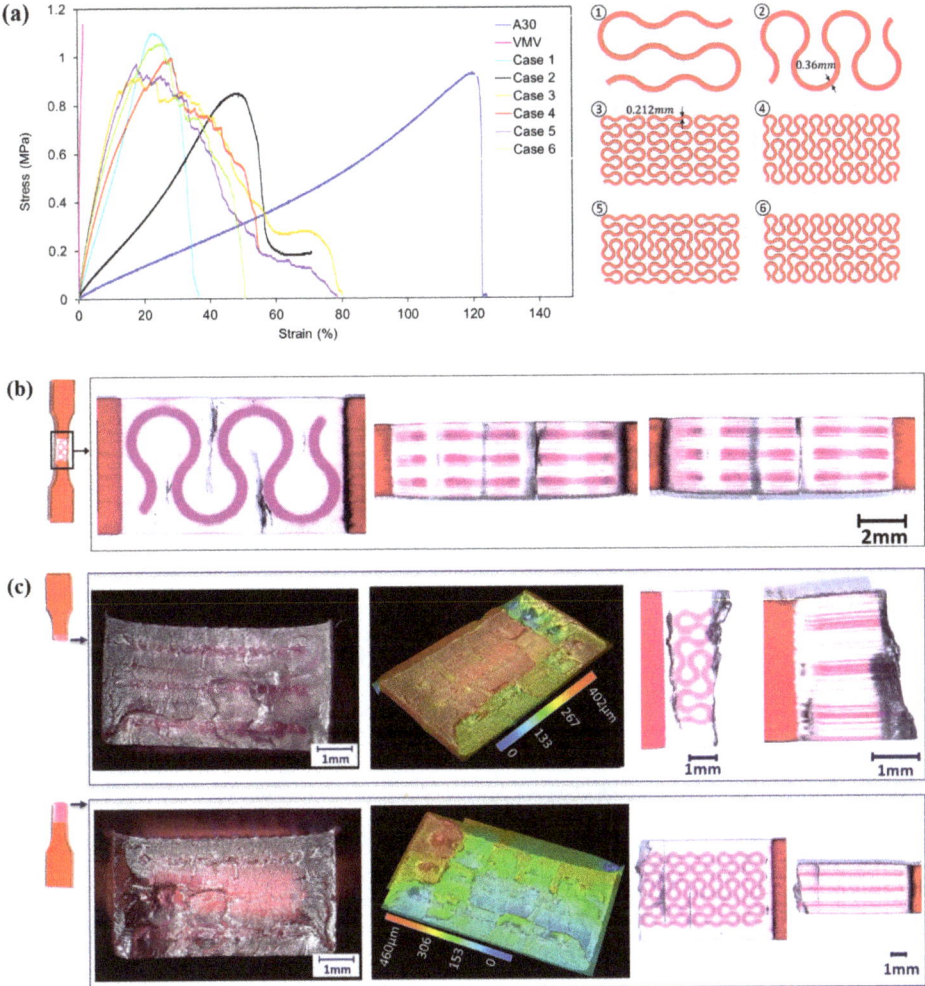

Figure 11. (**a**) Comparisons of tensile stress-strain curves of homogenous A30 (Agilus30), VMV (VeroMagentaV), and six hybrid designs (plan views of gauge sections for different hybrid cases are shown on the right); (**b**) microscopic images (50× magnification) taken from the top and side views of the case 2 specimen, showing the cracks happening in A30 stopping near VMV reinforcement; (**c**) microscope images (50× magnification) taken from fracture surfaces, top and side views of case 5 specimen, showing the uneven fracture surfaces, crack initiating points, and crack distributions.

With regard to ultimate tensile strength (UTS), Table 2 reveals that there is no obvious improvement in hybrid materials compared to homogenous A30. Particularly, the reinforcements in case two and case four contribute negatively to UTS. This phenomenon is attributed to the reinforcement along the tensile force direction that confines the transverse deformation of A30 in the gauge section. Hybrid case one exhibits the highest tensile strength, whereas all specimens fail at the cross-section without any reinforcement. The results indicate that the coupon sample design of case one could not transfer the stress from the A30 matrix to VMV reinforcement effectively. Moreover, experimental results are lower than the theoretical estimations due to manufacturing defects in samples. Even though the improvement in tensile strength is not remarkable by the inclusion of reinforcement, a clear upward trend of UTS is identified with the increasing ratio of reinforcements parallel to and perpendicular to the loading direction.

Post-mortem analysis of tensile samples was conducted using an optical microscope. Fracture surfaces, top, bottom, and side views of failed samples were studied to understand the crack propagation and failure patterns.

Representative microscope images are shown in Figure 11b,c. As we can see from Figure 11b, cracks happen in A30 and stop near the VMV reinforcement in one of the case two specimens. It is a result of the higher stiffness and strength of VMV than that of A30. The digital microscope images (Figure 11c) exhibit the uneven fracture surfaces of hybrid structures, which are captured for all other specimens as well. As revealed by the top view of the bottom half specimen (Figure 11c), small black lines and dots could be observed near the right extension. These are identified as the crack initiation points, which are caused by the stress concentration from the curved design of the Peano reinforcement. Then, the cracks propagate in the A30 matrix and form into a continuous crack, such as the long wave-shape black line shown in the top half of the specimen in Figure 11c. Additionally, no obvious delamination is captured at the A30/VMV interface, which indicates a reliable combining of the two different materials.

3.2. Compression Test Results and Discussions

The results obtained from compressive tests are summarised in Figure 12, with comparisons made in three loading directions and among different hybrid materials.

Generally, the stress developed in all VMV reinforced hybrid samples is remarkably higher than that in homogenous A30 samples according to Figure 12a–c. The result indicates that the inclusion of the VMV Peano curve strengthens the A30 matrix regardless of loading directions. With regards to different reinforcement hierarchies, it could be observed that the composite materials with the first order reinforcement (case one) yield a higher compressive strength than the second order materials (case 2–4). This phenomenon ascribes to a larger diameter of reinforcement in the hybrid case one design (2.8 mm) than the other cases (1.56 mm), given the same volume fraction of 10% for all. It is also worth noticing that the responses of case two, case three, and case four are relatively similar for all three loading directions. It is caused by their similar amount of reinforcement at the cross-section perpendicular to the compression force.

As compressive cubes are designed anisotropic, compressive properties of four different hybrid cases are studied in different directions as shown in Figure 12d–g. The results elucidate that all hybrid cubes exhibit the lowest compressive strength subjected to loading direction one. Since Peano reinforcements lie in five-layers perpendicular to the compressive loading direction one, the amount of VMV material in the corresponding cross-section is the least among all three directions. For hybrid case one, the highest compressive strength (5.55 MPa) is captured in loading direction two and the second highest is found in loading direction three (3.68 MPa). This can be explained by the amount of reinforcement along the loading directions, which restrains the transverse expansion of A30 and thereby increases the strength. For the other three hybrid cases with second order reinforcement, there is only a slight difference between the compressive strength in loading direction two and loading direction three. As the hierarchy of reinforcement increases from first order to second order, the amount in the difference of reinforcement both lying along or perpendicular to loading direction two and loading direction three becomes very small.

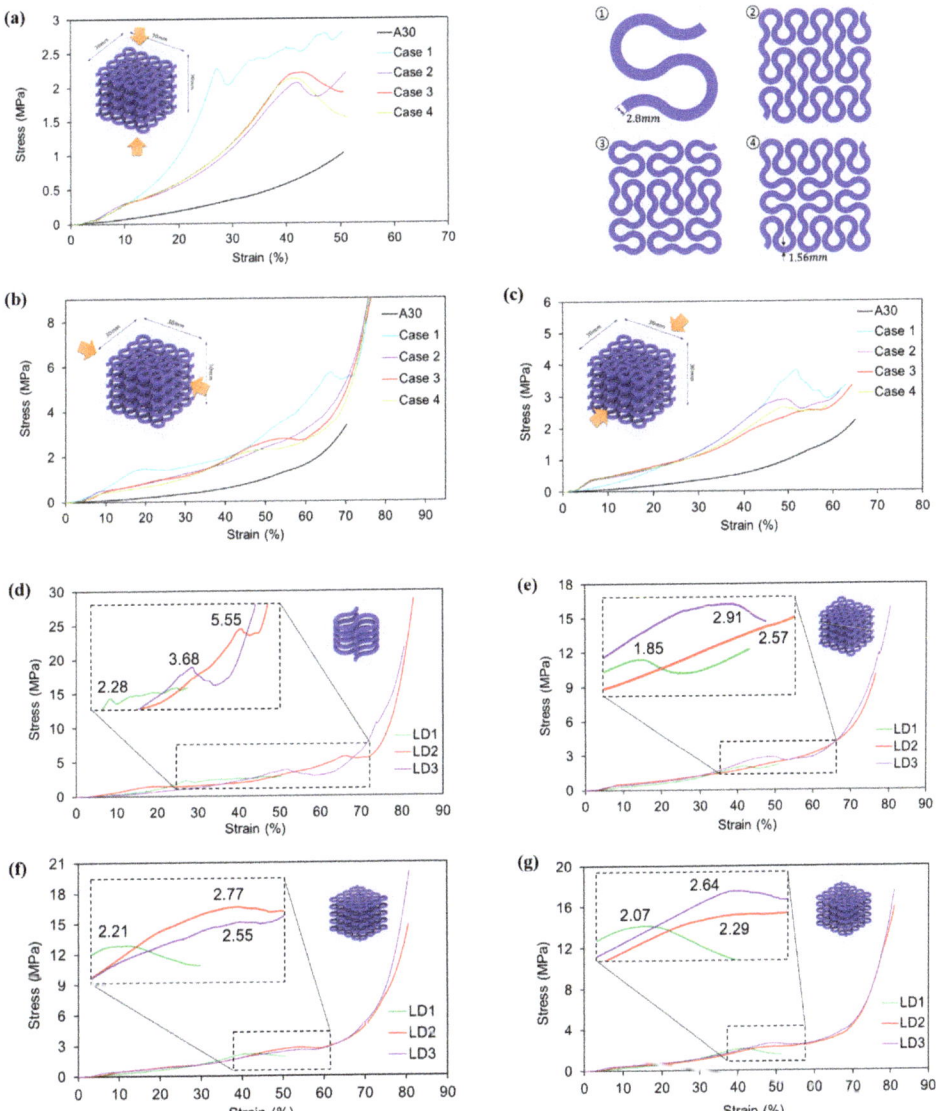

Figure 12. *Cont.*

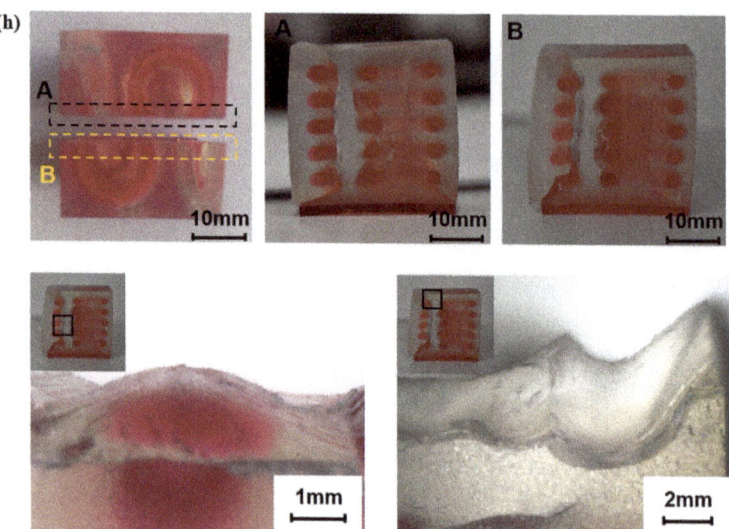

Figure 12. Stress-strain curves of different material designs, obtained from uniaxial compressive testings from (**a**) loading direction one (LD1); (**b**) loading direction two (LD2); (**c**) loading direction three (LD3). Comparisons of compressive stress-strain curves from three different loading directions for (**d**) design case one (pure horizontal/vertical first order); (**e**) design case two (pure horizontal/vertical second order); (**f**) design case three (mostly horizontal second order); (**g**) design case four (mostly vertical second order). (**h**) high resolution images (first row) and microscope images (bottom row) of case one, hybrid cube (pure horizontal/vertical first order) after compression from loading direction one. High resolution images show the cross-section A and B, from which cracks were found in the A30 matrix rather than A30/VMV interface. Microscope images show uneven fracture surfaces.

To investigate the failure pattern of compression cubes, high resolution pictures and optical microscope images are taken to capture the fracture surfaces of failed samples. As compression samples are not broken into pieces, a bandsaw is used to cut the failed specimens in half along the compressive loading directions. Figure 12h shows the fracture surfaces of the hybrid case one specimen after compression from loading direction one. The high-resolution images of cross-section A and B, in the top row, clearly show the existence of wavy cracks in the A30 matrix. Supportive information is provided by the microscope images of the fracture surfaces (bottom row of Figure 12h). Transparent A30 is observed on top of VMV (bottom left image in Figure 12h), indicating the existence of a crack in the matrix rather than any debonding of A30/VMV. The results imply that the interface between two different materials is relatively strong. Furthermore, a concave surface is captured in A30 after the compression (bottom right image in Figure 12h). The reason behind this phenomenon is the same as the wave-shape cracks observed in tensile samples. To be more specific, it is caused by the stress concentration in the A30 matrix due to the curved design of VMV reinforcement.

4. Conclusions

In this paper, we designed novel composite materials inspired by the Peano curve. PolyJet 3D printing technology was used to fabricate samples with Agilus30 (A30) and Vero-MagentaV (VMV). Mechanical properties were evaluated by mechanical tests, analytical predictions, and optical microscopy. Herein, the following conclusions are made:

- Compared to homogenous A30, all the hybrid tensile samples reinforced with VMV Peano curves yielded higher stiffness. This was attributed to the higher elastic modulus of VMV compared to A30. Consistent with the hypothesis, the elastic moduli

obtained from tensile tests were within the range approximated from the rule of mixture (ROM) for composites.
- Hybrid tensile samples, which were designed with the second order Peano reinforcement, generally had a higher elastic modulus than tensile samples with the first order Peano reinforcement. It can be concluded that the second order reinforcement designs were more effective than the first order ones in terms of stiffness enhancement. For the hybrid tensile designs with the same reinforcement hierarchy, the pure horizontal alignment of reinforcement always provided a higher stiffness than the pure vertical designs owing to a higher ratio of reinforcement parallel to the tensile force.
- Regarding ultimate tensile strength, the improvement of hybrid designs compared to homogenous A30 was not obvious. Hierarchy and alignment of Peano reinforcements seemed to have little influence on the tensile strength as the stress could not be transferred effectively from matrix to reinforcement. However, an increasing trend of UTS could be witnessed with the growing ratio of reinforcements parallel to, and perpendicular to the loading direction. Experimental results were much lower than theoretical predictions due to the 3D manufacturing defects.
- The introduction of VMV Peano reinforcement in the A30 matrix resulted in higher stiffness and strength of the compression cubes. The first order reinforcement exhibited the best performance in all three directions among four different designs. The responses of three different second order designs were similar under compression.
- The second order compression cubes exhibited similar properties in loading direction two and loading direction three, due to the similar amount of reinforcement in all three cases along the compressive force.

Author Contributions: C.W.: Data curation, Writing—original draft; T.T.D.: Resource, Writing—review & editing; P.T.: Conceptualization, Methodology, Review, Editing and Supervision. All authors have read and agreed to the published version of the manuscript.

Funding: This research received no external funding.

Data Availability Statement: Not applicable.

Acknowledgments: The authors acknowledge the facilities, and the scientific and technical assistance from RMIT Advanced Manufacturing Precinct.

Conflicts of Interest: The authors declare no conflict of interest.

References

1. Mandelbort, B.B. *Fractals: Form, Chance and Dimension*; American Institute of Physics: San Francisco, CA, USA, 1977.
2. Mandelbrot, B.B. Self-affine fractal sets, I: The basic fractal dimensions. In *Fractals in Physics*; Elsevier: Amsterdam, The Netherlands, 1986; pp. 3–15.
3. Hutchinson, J.E. Fractals and self similarity. *Univ. Math. J.* **1981**, *30*, 713–747. [CrossRef]
4. Gil'Mutdinov, A.K.; Ushakov, P.A.; El-Khazali, R. *Fractal Elements and their Applications*; Springer: Berlin/Heidelberg, Germany, 2017.
5. Puente, B.C.; Pous, A.R.; Romeu, R.J.; Garcia, F.X. Antenas Fractales o Multifractales. Spanish Patent 2,112,163, 19 May 1995.
6. Puente, C.; Romeu, J.; Pous, R.; Garcia, X.; Benitez, F. Fractal multiband antenna based on the Sierpinski gasket. *Electron. Lett.* **1996**, *32*, 1–2. [CrossRef]
7. Puente, C.; Romeu, J.; Cardama, R. On the behavior of the Sierpinski multiband antenna. *IEEE Trans. Antennas Propag.* **1998**, *46*, 517–524. [CrossRef]
8. Baliarda, C.P.; Romeu, J.; Cardama, A. The Koch monopole: A small fractal antenna. *IEEE Trans. Antennas Propag.* **2000**, *48*, 1773–1781. [CrossRef]
9. Yu, Z.W.; Wang, G.M.; Gao, X.J.; Lu, K. A novel small-size single patch microstrip antenna based on Koch and Sierpinski fractal-shapes. *Prog. Electromagn. Res. Lett.* **2010**, *17*, 95–103. [CrossRef]
10. Ali, J.K. A new microstrip-fed printed slot antenna based on Moore space-filling geometry. In Proceedings of the 2009 Loughborough Antennas & Propagation Conference, Loughborough, UK, 16–17 November 2009.
11. Cohen, N.; Hohlfeld, R.G. Fractal loops and the small loop approximation. *Commun. Q.* **1996**, *6*, 77–81.
12. Puente Baliarda, C. Fractal Antennas. Ph.D. Thesis, Technical University of Catalonia, Barcelona, Spain, 1997.
13. Kim, Y.; Jaggard, D.L. The fractal random array. *Proc. IEEE* **1986**, *74*, 1278–1280. [CrossRef]

14. McVay, J.; Hoorfar, A.; Engheta, N. Space-filling curve RFID tags. In Proceedings of the 2006 IEEE Radio and Wireless Symposium, San Diego, CA, USA, 17–19 October 2006.
15. McVay, J.; Engheta, N.; Hoorfar, A. Radio Frequeacy Identification Utilizing Passive Space-Filling Carves. In Proceedings of the USNC-URSI National Radio Science Meeting, Washington, DC, USA, 3–8 July 2005.
16. Fan, J.A.; Yeo, W.-H.; Su, Y.; Hattori, Y.; Lee, W.; Jung, S.-Y.; Zhang, Y.; Liu, Z.; Cheng, H.; Falgout, L.; et al. Fractal design concepts for stretchable electronics. *Nat. Commun.* **2014**, *5*, 3266. [CrossRef]
17. El-Khouly, E.; Ghali, H.; Khamis, S.A. High Directivity Antenna Using a Modified Peano Space-Filling Curve. *IEEE Antennas Wirel. Propag. Lett.* **2007**, *6*, 405–407. [CrossRef]
18. Zhu, J.; Hoorfar; Engheta. Peano antennas. *IEEE Antennas Wirel. Propag. Lett.* **2004**, *3*, 71–74.
19. Romeu, J.; Blanch, S. A three dimensional Hilbert antenna. In Proceedings of the IEEE Antennas and Propagation Society International Symposium (IEEE Cat. No.02CH37313), San Antonio, TX, USA, 16–21 June 2002.
20. Puente, C.; Romeu, J.; Pous, R.; Ramis, J.; Hijazo, A. Small but long Koch fractal monopole. *Electron. Lett.* **1998**, *34*, 9–10. [CrossRef]
21. Vinoy, K.; Jose, K.; Varadan, V. Multi-band characteristics and fractal dimension of dipole antennas with Koch curve geometry. In Proceedings of the IEEE Antennas and Propagation Society International Symposium (IEEE Cat. No.02CH37313), San Antonio, TX, USA, 16–21 June 2002.
22. Haji-Hashemi, M.R.; Moradian, M.; Mirmohammad-Sadeghi, H. Space-filling Patch Antennas with CPW Feed. *PIERS Online* **2006**, *2*, 69–73. [CrossRef]
23. Spence, T.G.; Werner, D.H. Werner, and propagation, Generalized space-filling Gosper curves and their ap-plication to the design of wideband modular planar antenna arrays. *IEEE Trans. Antennas Propag.* **2010**, *58*, 3931–3941. [CrossRef]
24. Dhar, S.; Ghatak, R.; Gupta, B.; Poddar, D.R. A Wideband Minkowski Fractal Dielectric Resonator Antenna. *IEEE Trans. Antennas Propag.* **2013**, *61*, 2895–2903. [CrossRef]
25. Abdul-Letif, A.M.; Habeeb MA, Z.; Jaafer, H.S. Performance characteristics of the Minkowski curve fractal antenna. *J. Eng. Appl. Sci.* **2006**, *1*, 323–328.
26. Piskun, V. Fractal Antenna Based on Peano-Gosper Curve. U.S. Patent 7,541,981, 2 June 2009.
27. Werner, D.; Kuhirun, W.; Werner, P. The Peano-Gosper fractal array. *IEEE Trans. Antennas Propag.* **2003**, *51*, 2063–2072. [CrossRef]
28. Werner, D.H.; Kuhirun, W.; Werner, P.L. Fractile Antenna Arrays and Methods for Producing a Fractile Antenna Array. U.S. Patent 7,057,559, 6 June 2006.
29. Chen, W.-L.; Wang, G.-M.; Zhang, C.-X. Small-Size Microstrip Patch Antennas Combining Koch and Sierpinski Fractal-Shapes. *IEEE Antennas Wirel. Propag. Lett.* **2008**, *7*, 738–741. [CrossRef]
30. Maza, A.; Cook, B.; Jabbour, G.; Shamim, A. Paper-based inkjet-printed ultra-wideband fractal antennas. *IET Microw. Antennas Propag.* **2012**, *6*, 1366–1373. [CrossRef]
31. Bangi, I.S.; Sivia, J.S. Minkowski and Hilbert curves based hybrid fractal antenna for wireless applications. *AEU Int. J. Electron. Commun.* **2018**, *85*, 159–168. [CrossRef]
32. Fu, H.; Xu, S.; Xu, R.; Jiang, J.; Zhang, Y.; Rogers, J.A.; Huang, Y. Lateral buckling and mechanical stretchability of fractal interconnects partially bonded onto an elastomeric substrate. *Appl. Phys. Lett.* **2015**, *106*, 091902. [CrossRef]
33. Ma, Q.; Zhang, Y. Mechanics of fractal-inspired horseshoe microstructures for applications in stretchable electronics. *J. Appl. Mech.* **2016**, *83*, 111008. [CrossRef]
34. Alcheikh, N.; Shaikh, S.F.; Hussain, M. Ultra-stretchable Archimedean interconnects for stretchable electronics. *Extrem. Mech. Lett.* **2018**, *24*, 6–13. [CrossRef]
35. Zhang, Y.; Fu, H.; Xu, S.; Fan, J.A.; Hwang, K.C.; Jiang, J.; Rogers, J.A.; Huang, Y. A hierarchical computational model for stretchable interconnects with fractal-inspired designs. *J. Mech. Phys. Solids* **2014**, *72*, 115–130. [CrossRef]
36. Huiskes, R.; Ruimerman, R.; Van Lenthe, G.H.; Janssen, J.D. Effects of mechanical forces on maintenance and adaptation of form in trabecular bone. *Nature* **2000**, *405*, 704–706. [CrossRef]
37. Farr, R.S. Fractal design for efficient brittle plates under gentle pressure loading. *Phys. Rev. E* **2007**, *76*, 046601. [CrossRef]
38. Farr, R.S. Fractal design for an efficient shell strut under gentle compressive loading. *Phys. Rev. E* **2007**, *76*, 056608. [CrossRef]
39. Farr, R.; Mao, Y. Fractal space frames and metamaterials for high mechanical efficiency. *EPL Europhys. Lett.* **2008**, *84*, 14001. [CrossRef]
40. Fan, H.; Jin, F.; Fang, D. Mechanical properties of hierarchical cellular materials. Part I: Analysis. *Compos. Sci. Technol.* **2008**, *68*, 3380–3387. [CrossRef]
41. Chen, Q.; Pugno, N.M. In-plane elastic buckling of hierarchical honeycomb materials. *Eur. J. Mech. A Solids* **2012**, *34*, 120–129. [CrossRef]
42. Sun, Y.; Pugno, N.M. In plane stiffness of multifunctional hierarchical honeycombs with negative Poisson's ratio sub-structures. *Compos. Struct.* **2013**, *106*, 681–689. [CrossRef]
43. Sun, Y.; Chen, Q.; Pugno, N. Elastic and transport properties of the tailorable multifunctional hierarchical honeycombs. *Compos. Struct.* **2014**, *107*, 698–710. [CrossRef]
44. Ajdari, A.; Jahromi, B.H.; Papadopoulos, J.; Nayeb-Hashemi, H.; Vaziri, A. Hierarchical honeycombs with tailorable properties. *Int. J. Solids Struct.* **2012**, *49*, 1413–1419. [CrossRef]

45. Haghpanah, B.; Oftadeh, R.; Papadopoulos, J.; Vaziri, A. Self-similar hierarchical honeycombs. *Proc. R. Soc. A: Math. Phys. Eng. Sci.* **2013**, *469*, 20130022. [CrossRef]
46. Oftadeh, R.; Haghpanah, B.; Papadopoulos, J.; Hamouda, A.M.; Nayeb-Hashemi, H.; Vaziri, A. Mechanics of anisotropic hierarchical honeycombs. *Int. J. Mech. Sci.* **2014**, *81*, 126–136. [CrossRef]
47. Zhang, Y.; Lu, M.; Wang, C.-H.; Sun, G.; Li, G. Out-of-plane crashworthiness of bio-inspired self-similar regular hierarchical honeycombs. *Compos. Struct.* **2016**, *144*, 1–13. [CrossRef]
48. Zhang, Y.; Wang, J.; Wang, C.-H.; Zeng, Y.; Chen, T. Crashworthiness of bionic fractal hierarchical structures. *Mater. Des.* **2018**, *158*, 147–159. [CrossRef]
49. Zhang, D.; Fei, Q.; Jiang, D.; Li, Y. Numerical and analytical investigation on crushing of fractal-like honeycombs with self-similar hierarchy. *Compos. Struct.* **2018**, *192*, 289–299. [CrossRef]
50. Meza, L.; Zelhofer, A.J.; Clarke, N.; Mateos, A.J.; Kochmann, D.; Greer, J.R. Resilient 3D hierarchical architected metamaterials. *Proc. Natl. Acad. Sci. USA* **2015**, *112*, 11502–11507. [CrossRef] [PubMed]
51. Wang, J.; Zhang, Y.; He, N.; Wang, C.-H. Crashworthiness behavior of Koch fractal structures. *Mater. Des.* **2018**, *144*, 229–244. [CrossRef]
52. Li, Y.; Ortiz, C.; Boyce, M.C. Bioinspired, mechanical, deterministic fractal model for hierarchical suture joints. *Phys. Rev. E* **2012**, *85*, 031901. [CrossRef] [PubMed]
53. Khoshhesab, M.M.; Li, Y. Mechanical behavior of 3D printed biomimetic Koch fractal contact and interlocking. *Extrem. Mech. Lett.* **2018**, *24*, 58–65. [CrossRef]
54. Dattelbaum, D.M.; Ionita, A.; Patterson, B.M.; Branch, B.A.; Kuettner, L. Shockwave dissipation by interface-dominated porous structures. *AIP Adv.* **2020**, *10*, 075016. [CrossRef]
55. Li, F.; Macdonald, N.P.; Guijt, R.M.; Breadmore, M.C. Increasing the functionalities of 3D printed microchemical devices by single material, multimaterial, and print-pause-print 3D printing. *Lab Chip* **2018**, *19*, 35–49. [CrossRef]
56. Zhang, X.; Wang, J.; Liu, T. 3D printing of polycaprolactone-based composites with diversely tunable mechanical gradients via multi-material fused deposition modeling. *Compos. Commun.* **2020**, *23*, 100600. [CrossRef]
57. Skylar-Scott, M.A.; Mueller, J.; Visser, C.W.; Lewis, J.A. Voxelated soft matter via multimaterial multinozzle 3D printing. *Nature* **2019**, *575*, 330–335. [CrossRef]
58. Yuan, C.; Wang, F.; Qi, B.; Ding, Z.; Rosen, D.W.; Ge, Q. 3D printing of multi-material composites with tunable shape memory behavior. *Mater. Des.* **2020**, *193*, 108785. [CrossRef]
59. Haneef, M.; Rahman, J.F.; Yunus, M.; Zameer, S.; Patil, S.; Yezdani, T. Hybrid polymer matrix composites for biomedical applications. *Int. J. Modern. Eng. Res.* **2013**, *3*, 970–979.
60. Naskar, A.K.; Keum, J.; Boeman, R.G. Polymer matrix nanocomposites for automotive structural components. *Nat. Nanotechnol.* **2016**, *11*, 1026–1030. [CrossRef] [PubMed]
61. Balakrishnan, P.; John, M.J.; Pothen, L.; Sreekala, M.S.; Thomas, S. Natural Fibre and Polymer Matrix Composites and Their Applications in Aerospace Engineering. In *Advanced Composite Materials for Aerospace Engineering*; Woodhead Publishing: Sawston, UK, 2016.
62. Tee, Y.L.; Peng, C.; Pille, P.; Leary, M.; Tran, P. PolyJet 3D Printing of Composite Materials: Experimental and Modelling Approach. *JOM* **2020**, *72*, 1105–1117. [CrossRef]
63. Lumpe, T.S.; Mueller, J.; Shea, K. Tensile properties of multi-material interfaces in 3D printed parts. *Mater. Des.* **2018**, *162*, 1–9. [CrossRef]
64. Soe, S.P.; Martindale, N.; Constantinou, C.; Robinson, M. Mechanical characterisation of Duraform® Flex for FEA hyperelastic material modelling. *Polym. Test.* **2014**, *34*, 103–112. [CrossRef]

Article

New Methodology for Evaluating Surface Quality of Experimental Aerodynamic Models Manufactured by Polymer Jetting Additive Manufacturing

Razvan Udroiu

Department of Manufacturing Engineering, Transilvania University of Brasov, 29 Eroilor Boulevard, 500036 Brasov, Romania; udroiu.r@unitbv.ro; Tel.: +40-268-421-318

Abstract: The additive manufacturing (AM) applications have attracted a great deal of interest with regard to experimental aerodynamic studies. There is a need for a universal roughness scale that characterizes different materials used in aerodynamic research. The main purpose of this paper is identification of the potential of a material jetting AM process to produce accurate aerodynamic surfaces. A new methodology to evaluate the roughness of aerodynamic profiles (airfoils) was proposed. A very short-span wing artifact for preliminary tests and a long-span wing model were proposed for design of experiments. Different artifacts orientations were analyzed, maintaining the same surface quality on the upper and lower surface of the wing. A translucent polymeric resin was used for samples manufacturing by polymer jetting (PolyJet) technology. The effects of main factors on the surface roughness of the wing were investigated using the statistical design of experiments. Three interest locations, meaning the leading-edge, central, and trailing-edge zones, on the upper and lower surfaces of the airfoil were considered. The best results were obtained for a sample oriented at XY on the build platform, in matte finish type, with a mean Ra roughness in the range of 2 to 3.5 µm. Microscopy studies were performed to analyze and characterize the surfaces of the wing samples on their different zones.

Keywords: additive manufacturing; polymers; material jetting; 3D printing; airfoil; aerodynamic model; design of experiments; surface roughness

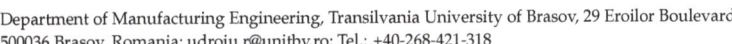

1. Introduction

Additive manufacturing (AM), known also as 3D printing, represents a key technology in the implementation of Industry 4.0 [1], based on its ability to fabricate highly complex and lightweight components directly from computer-aided design (CAD) files, saving time, cost, and effort. Additive manufacturing's applications have attracted interest within many fields, such as the transportation industry, health sector, energy sector, and consumer production.

Seven categories of AM processes are defined by ISO/ASTM 52900-15 [2] standard based on the different joining techniques of materials to make parts from 3D model data, as follows: vat photo-polymerization (VP), binder jetting (BJ), material extrusion (ME), material jetting (MJ), sheet lamination (SL), powder bed fusion (PBF), and directed energy deposition (DED). One of the most accurate AM processes is MJ [3]. Material jetting processes, which include polymer jetting (PolyJet) and multi-jet printing (MJM) technologies, can be defined as a technique that selectively deposits droplets of material and cured them onto a build platform.

The main materials types used in the seven individual AM processes described by AM standards are polymers, ceramics, metals, and composite materials. Polymers became very popular in AM being used in the most of the AM processes and targeting a variety of applications [4].

The AM applications for experimental aerodynamic studies have attracted much interest within the aerospace, automotive, and wind energy sectors. Thus, the aerodynamic parts obtained by AM are used for testing in a wind tunnel or as final components for UAVs (unmanned aerial vehicle), drones, wind turbines, and small aircrafts. The main requirements of an aerodynamic part are the light weight, a smooth surface, and good mechanical characteristics.

Surface roughness is an important factor in aerodynamics that can significantly influence the fluid dynamics and the heat transfer [5]. The roughness of the wing skin increases the skin friction drag, which is one of the parasite drag components [6]. Three factors cause the parasite drag of an aerodynamic vehicle (e.g., an aircraft): the aircraft's shape, construction type, and material. Parasite drag is split into three types: form drag, interference drag, and skin friction drag. The skin friction coefficients are sometimes based on experimental data for flat plates with various amounts of roughness. An inhomogeneous surface roughness distribution on an unmanned aircraft wing after many hours of flight was determined in [7]. It was mentioned that the initial roughness of the wing manufactured by a classical method not by additive manufacturing was 2 µm. The anisotropic influence of the winds during the flight, over the wing geometry, and the interferences between fuselage and wing were factors that increased the roughness.

Preliminary studies about additive manufacturing by the PolyJet process of airfoils for aerodynamic tests were performed in [8], but the quality of the airfoil surface was not investigated. The experimental coefficient of lift and drag of a NACA 2412 airfoil made by selective laser sintering (SLS) technology was studied in [9], but the surface roughness study was not carried out. Olasek et al. [10] evaluated a symmetrical NACA0018 airfoil model made by different materials and 3D-printing methods and concluded that surface roughness influences the aerodynamics characteristics of the airfoil. They also mentioned that the surface roughness is low for multi-jet modeling (MJM), moderate for SLS, and high for fused deposition modeling (FDM), but a range of roughness values were not mentioned. The rotor blades of a wind turbine rotor 3D printed by FDM technology were tested in wind tunnel by [11], but the surface roughness characterization was not performed. A UAV model was developed and manufactured using the binder jetting process by Junka et al. [12] for wind tunnel testing. This aerodynamic model was built by plastic powder and binder and then post-processed in order to obtain a good quality surface, but roughness investigation was not performed. These works demonstrated that 3D printing significantly changes the approach to experimental aerodynamics.

Three main tasks are significant to evaluate AM systems and processes for standardization and implementation in the industry: the performance characterization of the AM processes, AM part characterization, and AM system capability [3]. The main test methods for AM parts characterization are focused on mechanical properties [13–15], surface aspects [16,17], and dimensional geometry requirements [18]. Based on artifacts or customized models [19–21], the performance of the AM process can be investigated. The basic characterization of an AM system can be achieved via geometric accuracy [20], surface finish [21], and minimum feature sizes of the artifact [19]. The standards related to AM do not dictate a specific measurement method of artifacts features [22].

The surface quality of the AM parts was investigated in many studies mainly focused on the surface roughness determination [21,23–25]. The main factor that affects the surface roughness in different AM processes is the deposition layer thickness. It was reported based on experimental study that components generated through material jetting technology have superior surface quality than material extrusion components [26]. Additive manufacturing processes that use very thin layers deposition reduce the surface roughness, improving the surface quality. PolyJet technology, using deposition layers of 16 µm [27,28], significantly reduce the surface roughness of the parts. Part orientation on the built platform influences the surface roughness of the AM part [29]. Many studies have investigated it for different AM processes. An optimal part orientation achieves good results [30]. In addition, the surface roughness of the 3D-printed parts can be affected by external factors. Thus, a wearing analysis about PolyJet parts was performed in [31].

Some studies examined the material properties' characterization, process parameters, dimensional, and geometrical characterization of material jetting, but a lack of knowledge around the aerodynamic parts such as airfoils was found. There is also a need for a universal roughness scale that can describe every type of roughness for different materials used in aerodynamic studies. Thus, printing quality (surface roughness) is an important aspect when the parts are meant for the aerodynamic tests. Based on the AM standards, there is no general "best practice" to perform the measurements in AM, especially for aerodynamic surfaces (e.g., airfoils and wings).

The main aims of this article are to define a methodology for evaluating the surface quality of aerodynamic surfaces and to identify the potential of the material jetting AM process to produce accurate aerodynamic surfaces (e.g., airfoils and wings). A case study regarding polymer jetting process and its materials validates the proposed methodology.

2. Materials and Methods

The main objective of the proposed methodology is to evaluate the performance of an additive manufacturing process to produce aerodynamic artifacts. This methodology includes screening design of experiments (DOE) for a very short aerodynamic artifact and confirmatory experiments using a long aerodynamic artifact, which is followed by analysis and interpretation of the results, as is shown in Figure 1. The aerodynamic artifacts that were analyzed in this paper were airfoils and wings. Airfoil is a cross-sectional shape of an object whose motion through a fluid (e.g., air) is capable of generating significant lift force and a small drag force.

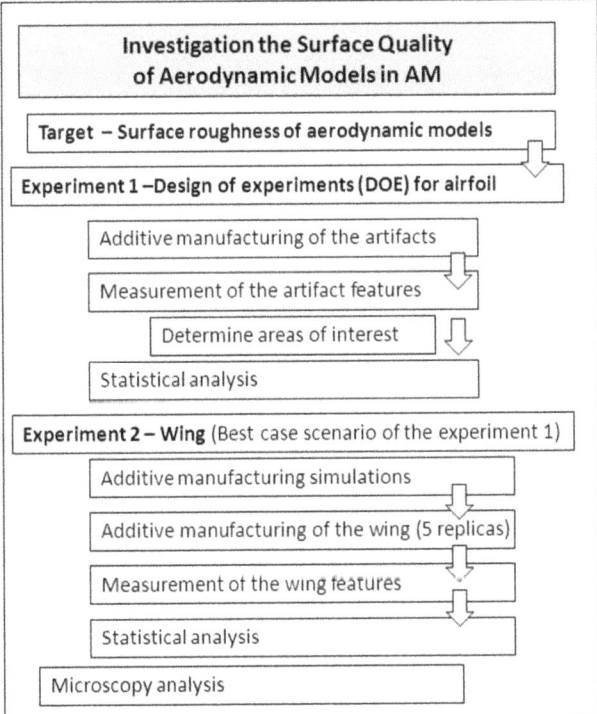

Figure 1. Flowchart of the proposed methodology of investigation of the surface quality of aerodynamic models produced by additive manufacturing.

The target of the experiments is the surface roughness of the aerodynamic models obtained by an AM process. From aerodynamic considerations, the surface roughness of the upper and lower surface of the airfoil should have similar values. Therefore, the airfoil should be optimally positioned on the built platform to achieve it. In addition, the orientation of the part on the build platform influences the printing time and materials consumption. These parameters influence the total price of the 3D-printed part.

2.1. Aerodynamic Artifacts, Design of Experiments, and Simulations

Two types of aerodynamic artifacts were designed using the SolidWorks version 2016 software (Dassault Systèmes, Waltham, MA, USA), a very short-span wing (VS-SW) model denoted airfoil and a long-span wing (L-SW) model denoted wing. Both aerodynamic artifacts are designed using an asymmetrical airfoil such as NACA 8410 airfoil, with a chord length of 85 mm and taper ratio of 1 (Figure 2). Some basic terms related to airfoil are upper curve, lower curve, and chord line, as shown in Figure 2. The aerodynamic artifacts are designed with a span of 10 mm for the VS-SW model (Figure 3) and 200 mm for the L-SW model (Figure 4), respectively. The main characteristics that allow defining the locations on the wing are wing lower and upper surfaces, inboard and outboard of the wing, and three distinct zones: the leading edge, central, and trailing edge (Figure 4).

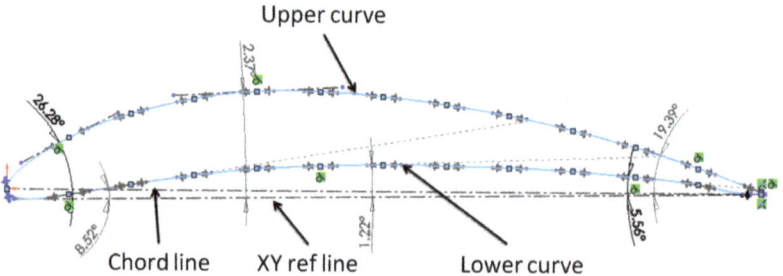

Figure 2. Terminology of asymmetrical airfoil curve—NACA 8410.

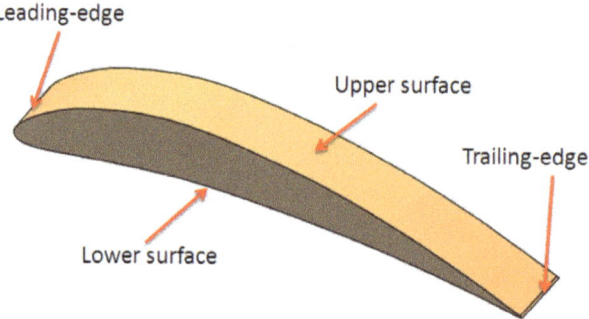

Figure 3. The very short-span wing model (airfoil).

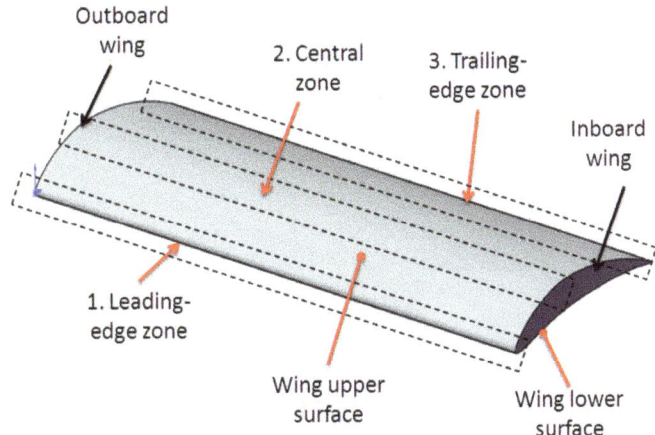

Figure 4. The long-span wing model.

The VS-SW artifact is used for preliminary tests as screening design of experiments. Screening design of experiments is used to reduce a large set of factors, and usually, multiple replicates are not used. If a prediction model is searched for, using multiple replicates can increase the precision of the model. In this case, usually a minimum five replicas are used. In addition, the resources can dictate the number of replicates if the experiment is costly.

The preliminary experiment about the surface roughness investigation of the airfoil was designed [32] by choosing the control factors that affect the surface quality and their levels. There are many factors that affect the surface roughness in additive manufacturing [21]. The selection of the control factors depends on the particularities of the additive manufacturing process and the artifacts' geometry. The following control factors that affect the surface quality were taken: airfoil orientation, surface finishing type, airfoil surface, and interest location. Details about the control factors and their levels are shown in Table 1.

Table 1. Control factors and their level.

Level	Target	Airfoil Orientation [1]		Surface-Finishing Type		Airfoil Surface		Interest Location	
	Symbol	Symbol	Value	Symbol	Value	Symbol	Value	Symbol	Value
1		1	XY	1	Matte	1	upper	1	Leading-edge zone
2	Ra	2	YX	2	Glossy	2	lower	2	Central zone
3		-	-	-	-	-	-	3	Trailing-edge zone

[1] Only the aerodynamic artifact orientations that allow obtaining a similar roughness distribution on the upper and lower surface of the airfoil were considered.

The airfoil orientation on the build platform is considered a two-level factor with basic orientations of parallel and perpendicular to the scanning direction (called XY and YX, respectively). The condition to keep the same surface quality on the upper and the lower surface of the airfoil was taken into consideration for airfoil orientations. The orientation of the AM build platform coordinate system was defined based on the ISO/ASTM 52921-13 standard [33]. The layout of artifact orientations on the build platform is shown in Figure 5.

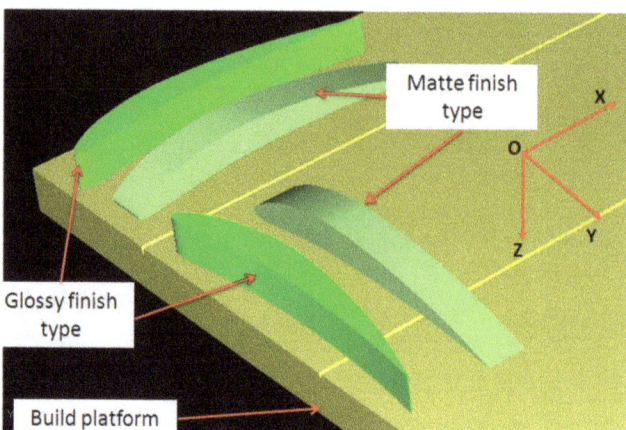

Figure 5. XY and YX orientations of the airfoil artifact. The first symbol indicates the longitudinal direction on the printing table and the second one indicates the transverse direction.

Three interest locations for surface quality (roughness) investigations, meaning leading-edge, central, and trailing-edge zones, on the upper and lower surfaces of the airfoil were proposed based on aerodynamics considerations.

One distinctive factor of PolyJet technology is the surface finishing type. Matte and glossy finishing are the levels of finish type. A thin layer of support material is applied around the surface of the part in matte finish printing. It allows obtaining a uniform surface of the part. In the case of the parts printed in glossy finish type, the support material is deposited only on the bottom surfaces of the part, and the upper surfaces are glossy.

A general full factorial design with 24 factor combinations was performed to be able to investigate the influence of the control factors (Table 1) on the surface roughness of the airfoil. A statistical analysis of the data was performed for the airfoil, investigating and characterizing the effects of control factors and their interactions on the surface roughness of the airfoil. The analysis of variance (ANOVA) approach using a generalized linear model (GLM) was used. The statistical analysis was performed using the Minitab 17 software (Coventry, UK) [34]. The statistical indicators p-value and F-value determined in the ANOVA table indicate the significance of the results.

Based on the best-case scenario obtained from preliminary tests, a confirmatory experiment about roughness investigation was performed using the L-SW artifact. The span of this artifact is larger than the span of the VS-WS artifact. This larger wingspan is a factor that influences the manufacturing time of the wing in different orientation on the build platform.

Thus, some simulations of the AM process are required in order to minimize the manufacturing time and material consumption. The building time and the quantity of the model and support material were determined by simulation in Objet Studio software (Stratasys, Rehovot, Israel), as shown in Table 2. From the analysis of the simulation, the following conclusions can be drawn:

- The lowest building times were obtained in the case of XY matte and XY glossy orientation. The lowest time was obtained in the case of XY glossy, but different quality of the upper and lower surface of the airfoil was observed based on the support material influence on the lower surface.
- A medium time was obtained in the case of YX matte and YX glossy orientation.
- For the ZX and ZY orientation, a high building time was obtained, the highest being obtained for the ZY orientation.

Table 2. Estimated 3D-printing parameters for different orientation on the build platform of the L-SW model.

Wing Orientation	Surface Finishing Type	Building Time (hour:min)	Model Consumption (g)	Support Consumption (g)
XY	matte	1:38	172	97
	glossy	1:34	170	81
YX	matte	3:16	177	102
	glossy	3:13	174	86
ZX	matte	10:59	187	99
	glossy	10:56	175	35
ZY	matte	22:46	215	127
	glossy	22:40	202	59

Four orientations of the wing on the build platform (Figure 6) resulted to be candidates that can be taken into consideration, keeping the same quality of the upper and the lower surface of the wing.

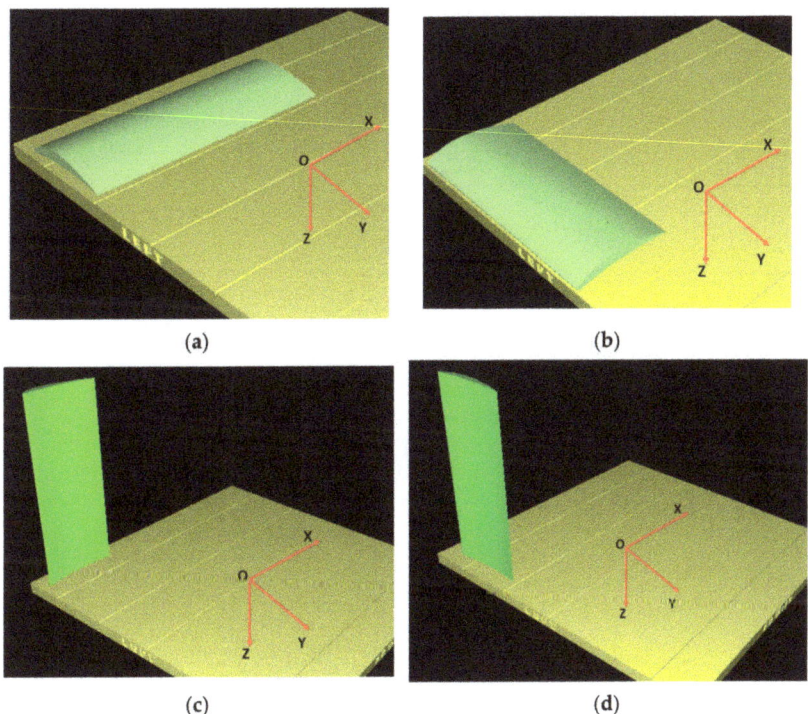

Figure 6. Simulations of the L-SW model in different orientation: (**a**) XY glossy; (**b**) YX matte; (**c**) ZX matte; (**d**) ZY matte.

Based on the simulations and the results obtained from the preliminary experiment (screening DOE), an optimal 3D-printing configuration for airfoils manufactured by Poly-Jet technology was determined to be XY matte. Five samples of wing in this optimal 3D-printing configuration were manufactured. The experimental surface roughness distribution on the long-span wing was analyzed.

2.2. Process Specification

All the samples (VS-SW and L-SW) were converted into standard triangle language (STL) files, imported into Objet Studio version 8.0.1.3 software (Stratasys, Rehovot, Israel), and manufactured using the Objet EDEN 350 PolyJet machine (Stratasys, Rehovot, Israel) [35]. The STL file conversion tolerances were set to a deviation of 0.01 mm and an angular tolerance of 4 degrees.

The materials known as FullCure 720 as model material and FullCure 705 as support material, supplied by Stratasys, were used to fabricate all samples. The composition of the Objet Fullcure 720 resin consists of acrylic monomer, urethane acrylate oligomer, epoxy acrylate, and photoinitiator. The main properties of Objet Fullcure 720 resin, also known as RGD720, are shown in Table 3 [36]. The support material, FullCure 705 resin, consists of acrylic monomer, polyethylene glycol 400, propane-1, 2-diol, glycerol, and photoinitiator. Diphenyl (2,4,6-trimethylbenzoyl) phosphine oxide is the photoinitiator used in UV treatment, as mentioned in [37]. The water contact angle of FullCure 720 material was investigated in [38], and it concluded that the Fullcure 720 is hydrophilic, with average contact angles of 81.0°. A Fourier transform infrared spectroscopy (FTIR) analysis of a material from the same acrylic family as FullCure720 was performed in [39]. They concluded that the spectrum shows that the material is acrylic based, (C=O) at 1721 cm^{-1}. The chemical and physical characterization of polymers used in the PolyJet process will be investigated in a future work. The polymers characterization should follow a route as mentioned in [40].

Table 3. Objet FullCure 720 properties [36].

Property	ASTM	Metric
Tensile Strength	D-638-03	50–60 MPa
Elongation at Break	D-638-05	15–25%
Flexural Strength	D-790-03	60–70 MPa
Rockwell Hardness	Scale M	73–76 Scale M
Water Absorption	D-570-98 24 h	1.5–2.2%
Polymerized Density	ASTM D792	1.18–1.19 g/cm^3

The Objet EDEN 350 PolyJet machine works on polymer jetting technology that is derived from drop-on-demand (DOD) inkjet technology [41]. Basically, the process consists of depositing layers of resins that are 0.016 mm thick, which are leveled by a roller and hardened by ultraviolet (UV) light. The main PolyJet process parameters were temperature of around 72° Celsius of the print heads and the photopolymer resins and a vacuum of 6.2 atm applied in the print heads. The experiments were performed under a controlled laboratory temperature of 20° Celsius and relative humidity of 30%.

Only the specimens printed in matte finishing (Figure 7) were post-processed by pressure water jet to remove the support material that surrounded the parts.

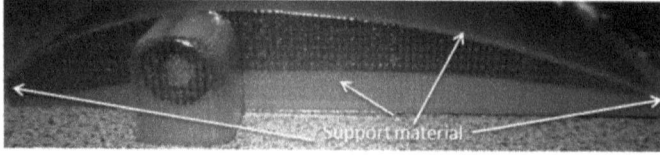

Figure 7. Specimen printed in matte finishing surrounded of support material.

2.3. New Measurement Strategy for Surface Roughness of Airfoils

The roughness measurement strategy for an airfoil includes two tasks: establish the measurement areas of interest and apply filters (i.e., the cut-off length). The filters were chosen based on DIN EN ISO 4288 standard [42]. The Gauss-filtered measurements were set up as follows: an evaluation length of 12.5 mm and a cut-off value of 2.5 mm.

A new roughness measurement strategy of the wing is proposed, which consists of evaluating the surface roughness in three interest locations meaning leading-edge, central, and trailing-edge zone, on upper and lower surfaces of the airfoil (Figure 8). The interest locations were denoted "A" for the leading-edge zone, "B" for the central zone, and "C" for the trailing-edge zone, as is shown in Figure 8. In the case of the long-span wing, three sections denoted 1, 2, and 3 placed along the span wing were taken into consideration for the roughness measurements. One measurement section was considered for the VS-SW artifacts.

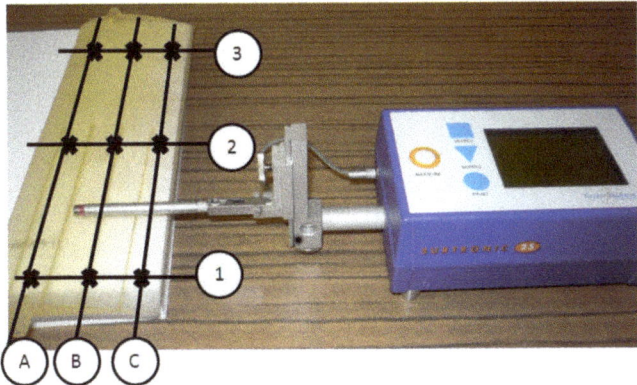

Figure 8. Measurement scheme of the airfoil roughness. Surface roughness measurement of the upper surfaces.

A Surtronic 25 contact surface roughness tester (Hoofddorp, The Netherlands) from Taylor Hobson was used to perform the measurements. The contact surface roughness tester was calibrated before performing the measurements. Profile measurements were repeated five times on each location of the airfoil surface, and the mean value was taken. The surface roughness Ra (the arithmetic mean deviation) was evaluated. The variability caused by the roughness measurement device was investigated, and data were processed within Minitab 17 software (Coventry, UK) using Gage R&R study [43].

A quality inspection through a microscopy study of the airfoils was performed using a Mitutoyo TM-1005 B optical-digital microscope (Mitutoyo, Kawasaki, Japan) with a digital micrometer head.

3. Results

The results of the performance of PolyJet process on the Objet EDEN 350 PolyJet machine to produce aerodynamic artifacts were analyzed taking into account the following considerations:

- The experimental surface roughness distribution on the upper and lower surface of the airfoil printed in two different quality modes and different orientations on the build platform;
- Surface quality issues of airfoil samples;
- The experimental analysis by microscopy of airfoils printed in different orientation;
- Results of statistical analysis.

All artifacts of the very short-span wing were manufactured in 2 h and 14 min, using 39 g of model material and 27 g of support material. The processing time for each long-span wing printed in XY orientation (best-case scenario) was 1 h and 38 min, and the consumption was 172 g of model material and 97 g of support material.

3.1. The Experimental Surface Roughness Analysis on Airfoils

Surface roughness distribution along the very short-span wing (airfoil) determined from experiments is shown in Figure 9.

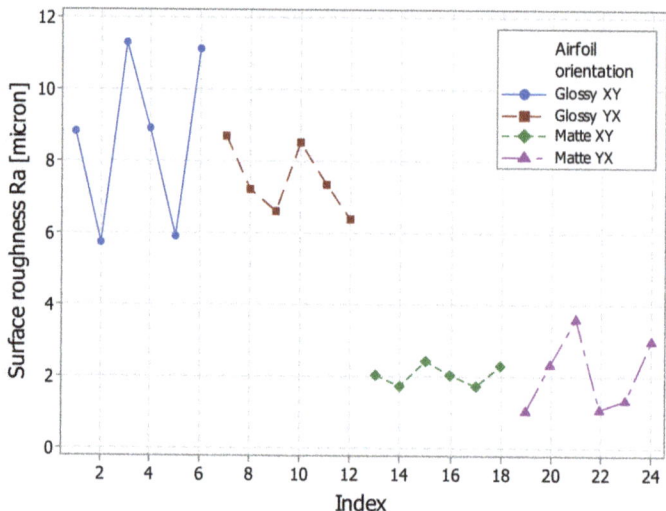

Figure 9. Surface roughness (Ra) distribution of the airfoils: glossy XY, glossy YX, matte XY, and matte YX.

The experimental roughness (Ra) values of the airfoil printed in matte finish were found in the range of 1.06 to 3.62 microns for the YX orientation and 1.74 to 2.46 microns for the XY orientation, as shown in Figure 9. The roughness of the artifacts printed in glossy finish presented higher values than the matte finish artifact, in the range of 5.72 to 11.3 microns for the XY orientation, and 6.4 to 8.68 microns for the YX orientations (Figure 9). Similar quality of the upper and lower surface of the airfoil was found in the mentioned range both for matte and glossy printed samples.

The lowest and relatively uniform roughness (Ra) was obtained for the airfoils printed in matte mode in the XY orientation. This is a reason why the L-SW experimental wings are printed in matte finishing mode.

3.2. Results of Statistical Analysis

The roughness tester variation based on the Gage R&R [43] study was much smaller than the variation of the surface roughness of the 3D-printed parts, proving the repeatability of the measurement system, as is mentioned also in [21].

From the ANOVA table (Table 4), the surface finishing type and the interest location are the significant factors that have a higher influence on the Ra roughness of the airfoils, taking into account their *p*-values. In addition, only the factor surface finishing type showed F_{exp} values greater than the critical F-value of 0.1% at $\alpha = 0.001$. Thus, the results were significant at the 0.1% significance level. The percentage contributions ratios for all the factors are presented in Table 4. The most significant factor on the roughness parameter (Ra) was the surface finishing type, which explained 82.86% of the total variation. The next contribution on Ra came from the interest location, with a contribution of 4.46%. The airfoil orientation and airfoil surface factors have no important effect on roughness parameter (Ra), which is based on a higher *p*-values and low percentage contribution (PC%).

Table 4. The percentage contribution ratio based on generalized linear model (GLM).

Source	DF	Seq SS	Seq MS	F_{exp}	$F_{0.1\%}$	p	PC (%)
Airfoil orientation	1	1.955	1.955	1.15	15.37	0.299	0.07%
Surface finishing type	1	214.503	214.503	125.66	15.37	<0.001	82.86%
Airfoil surface	1	0.137	0.137	0.08	15.37	0.781	0.005%
Interest location	2	11.548	5.774	3.38	10.38	0.057	4.46%
Error	18	30.726	1.707				11.86%
Total	23	258.869					100%

The evaluation of the influence of the control factors on the surface roughness (Ra) was performed through graphical analysis. The following graphs were obtained based on the statistical results, the main effects plot, interaction effects plot, and interval plot of Ra versus each factor.

The main effects for surface roughness were the airfoil orientation at level 1 (XY), the finish type at level 2 (glossy), and the interest location at level 3 (trailing-edge zone), as is shown in Figure 10. It is obvious that the factors surface finishing type, interest location, and their interaction had a significant influence on the surface roughness, as shown in Figures 10 and 11.

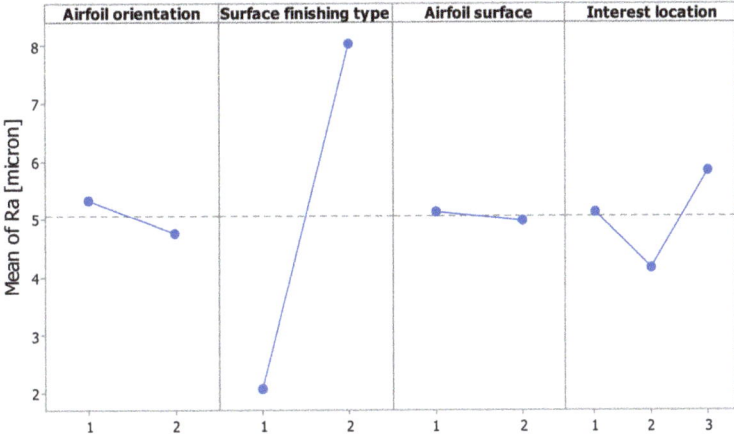

Figure 10. Main effects plot for surface roughness Ra.

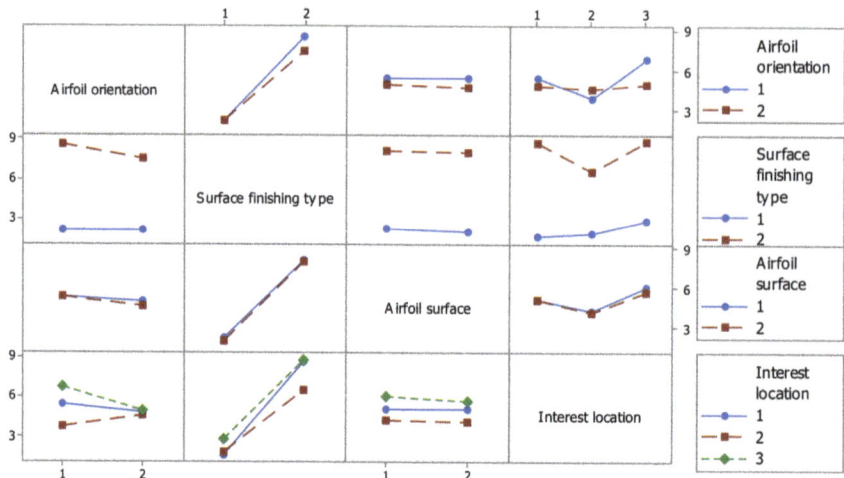

Figure 11. Interaction effects plot for surface roughness Ra.

The interval plots with standard error bars of each factor versus the roughness (Ra) are shown in the graphs from Figure 12. The difference between the means for Ra in the surface finishing type was significant because the interval bars did not overlap, as is shown in Figure 12a. While the means appear to be different, the differences for Ra in the airfoil orientation and airfoil surface were probably not significant because the interval bars easily overlapped (Figure 12b,d). The interest location (Figure 12c) had an influence on Ra, and it seems that at the central zone of the airfoil, the mean of Ra was lower, while for the leading-edge and trailing-edge zones, the mean was higher.

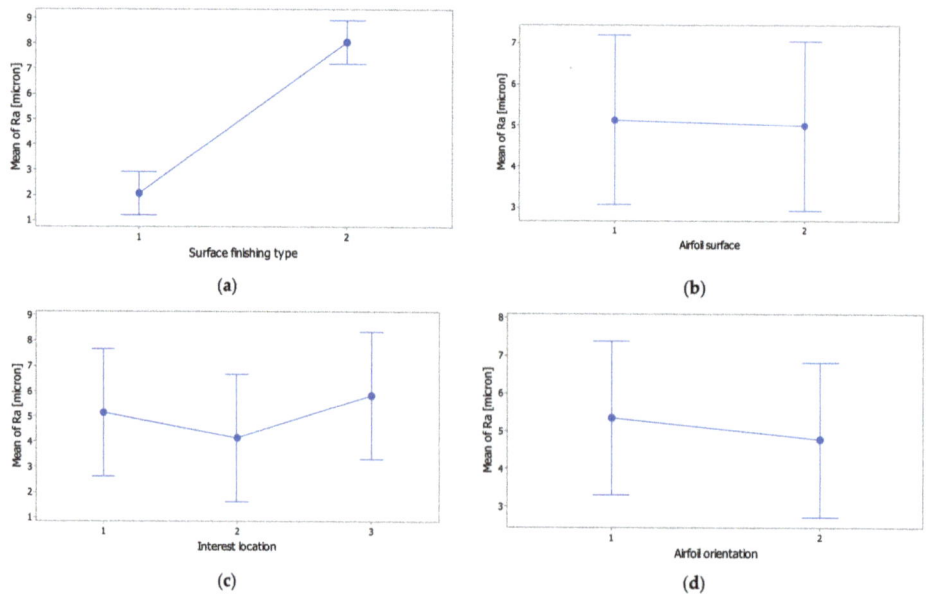

Figure 12. Individual standard deviations were used to calculate the intervals plot of surface roughness (Ra) versus (**a**) surface finishing type, (**b**) airfoil surface, (**c**) interest location, and (**d**) airfoil orientation; bars are standard errors of the mean.

3.3. Tests Results about Long-Span Wing

The results of the statistical analysis of the L-SW data show that the coefficients of variation for all the wing regions are lower than 10%, which assures the data heterogeneity and expresses the repeatability of the experiments, as shown in Table 5. The coefficient of variation (CV) is a measure of spread that describes the variation in the data relative to the mean. The standard error of the mean estimates the variability between samples, whereas the standard deviation measures the variability within a single sample.

Table 5. Statistics of wing surface roughness Ra for five samples.

Airfoil Region	Mean Roughness Ra [micron]	Standard Deviation [micron]	Coefficient of Variation [%]
A_upper_surf	2.47	0.225	9.08
B_upper_surf	1.86	0.089	4.82
C_upper_surf	2.22	0.131	5.9
A_lower_surf	2.43	0.221	9.09
B_lower_surf	1.80	0.090	5.02
C_lower_surf	2.22	0.094	4.25

The highest values for the surface roughness of the wing were found on the leading edge of the airfoil. This can be explained taking into accord that the angle between the wing surface and the horizontal plane is around 25°, as confirmed by the reference [21]. The smallest Ra values are found on the central zone of the wing, which could be considered a near-horizontal surface.

The interval plots of the surface roughness of the wing are in the range ±0.1 micron for all samples (Figure 13). Individual standard deviations were used to calculate the interval plot.

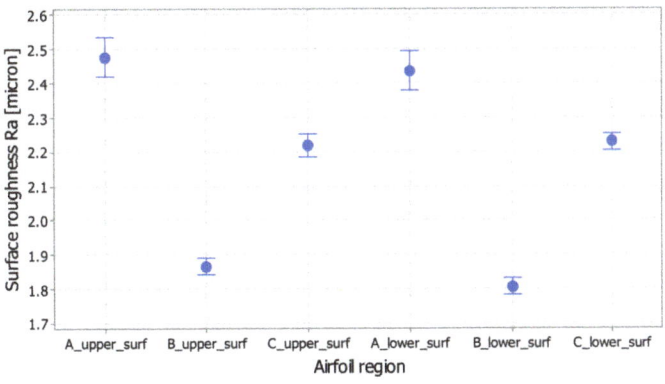

Figure 13. Interval plot of surface roughness for airfoil regions (matte XY orientation); bars are one standard error from the mean.

3.4. Quality Inspection of Airfoil Surface Based on Visual Inspection and Microscopy Analysis

Visual inspection and microscopy study were performed to analyze and characterize the surfaces of the samples in their different zones.

Surface errors on PolyJet-printed parts in glossy mode were determined in some studies [21,25]. Thus, rough surface areas were observed on the surface of the cylindrical parts printed on the Objet EDEN 350 perpendicular to the scanning direction [25]. In addition, horizontal steps marks [21] were visually observed on the flat faces oriented at 75° and 85° relative to the XOY plane. However, no visual flaws were detected on the vertical flat walls printed in the glossy style.

Based on visual inspection, two types of errors (rough surface areas) on the vertical walls of the airfoils (VS-SW) printed in glossy mode were determined (observed), as is shown in Figure 14. The first error consisting in vertical stripes on the airfoil surface was observed on the surface perpendicular on the scanning direction (X-axis direction) of the 3D printer. These were caused by the lower resolution of 0.042 mm in the X-direction and Y-direction compared to 0.016 mm in the Z-direction. The "vertical stripes" errors are predominant on the airfoil artifact printed in the XY orientation. The glossy sample printed in the YX direction presents a surface with a great density of points, which results in a homogeneous texture. There were no defects detected on the airfoils surface printed in matte mode, as is shown in Figure 15.

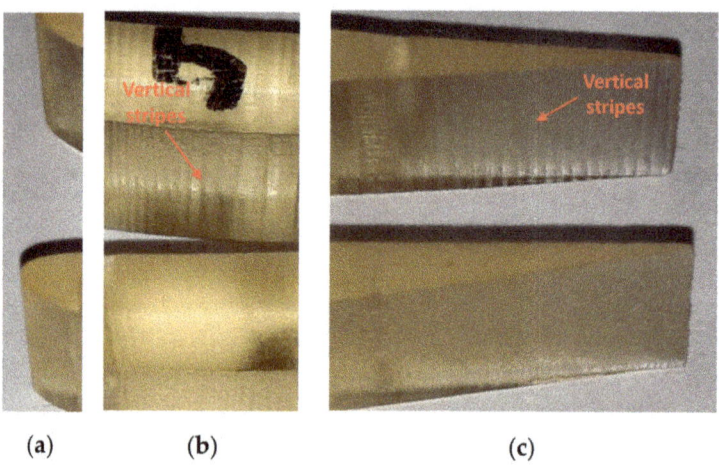

Figure 14. Comparative study of the airfoil surface in different interest locations for glossy XY (upper) and glossy YX (lower): (**a**) leading edge; (**b**) central zone; (**c**) trailing edge.

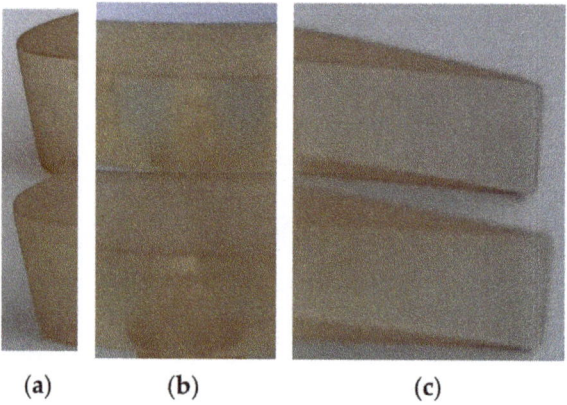

Figure 15. Comparative study of the airfoil surface in different interest locations for matte XY (upper) and matte YX (lower): (**a**) leading edge; (**b**) central zone; (**c**) trailing edge.

In the XY-direction orientation, fewer nozzles of the printer head are used compared to the YX orientation, as is shown in the partial views from Figure 16. Each nozzle deposits a train of droplets of resin grouped in a cylinder shape, and the level Z cylinders form a layer. Cylinders are similar to long fibers within composite materials. It can be seen that within the layers, the XY specimens have longer but fewer fibers than the YX specimens [14].

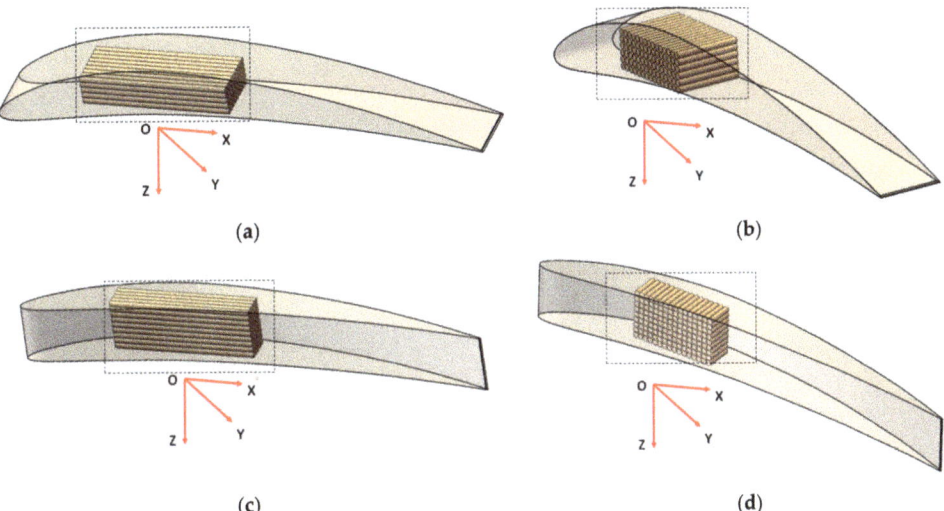

Figure 16. VS-SW specimens with partial views indicating the layers of cylinders for the different print orientations: (**a**) matte XY; (**b**) matte YX; (**c**) glossy XY; (**d**) glossy YX.

There are more intersections between cylinders from the layers and airfoil surface in the YX orientation. In addition, each intersection could be approximated as a circular shape of very small size, which confers a homogeneous texture, confirming the experimental observations (Figure 17b). In the case of XY orientations, the intersections between simulated layers (e.g., the lateral surface of the cylinders) and airfoil surface leads to straight vertical stripes (Figure 17a). These vertical stripes are more pronounced in the glossy printed mode based on an increased UV exposure. In the matte finishing, a theoretical 3D model is difficult to be drawn because, an additional support material layer is deposited on the airfoil surface, which allows obtaining a uniform texture.

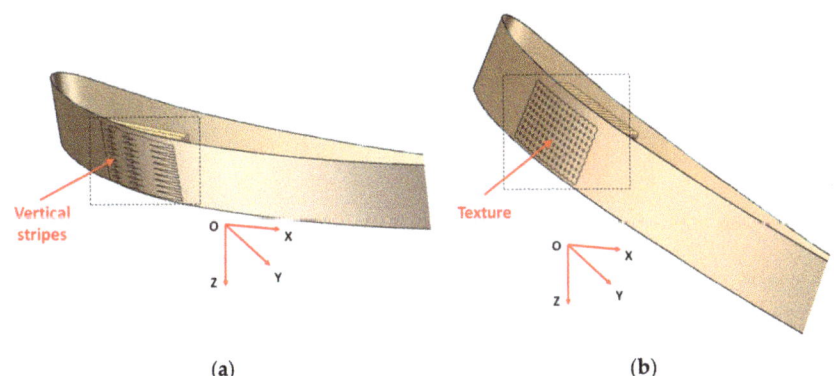

Figure 17. Theoretical texture of the airfoil surface: (**a**) glossy XY; (**b**) glossy YX.

The microscopy analysis study was performed for quality inspection of airfoil surfaces. All the surfaces of the matte specimens are affected by the material support. Very small inclusions of the support material were detected on the surface of these specimens even after post-processing by cleaning with a pressure water jet.

The texture of matte specimens is shown in Figure 18c,d. Specimens printed in glossy mode present different kind of texture depending on their orientation. There can be seen a homogeneous airfoil surface for the glossy YX specimen (Figure 18). The width values of vertical stripes (Figure 18a) detected on the glossy XY specimen were in the range 0.213 to 0.386 mm, which were measured within the microscopy analysis.

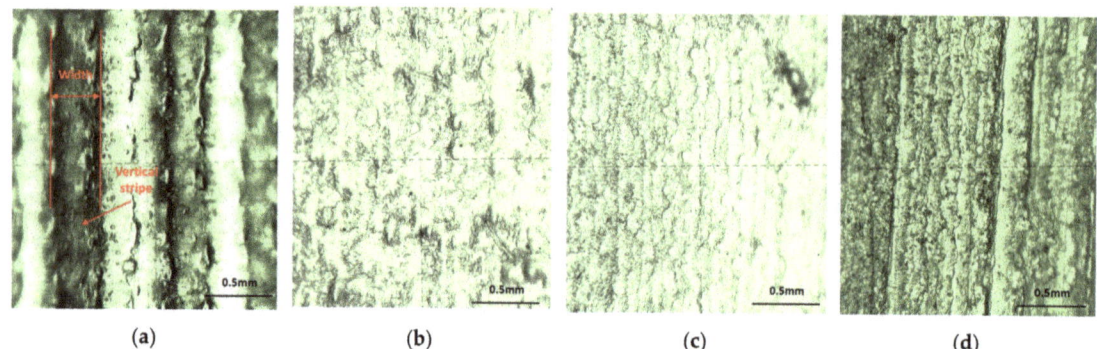

Figure 18. The texture of the airfoil surface: (**a**) glossy XY; (**b**) glossy YX; (**c**) matte XY; (**d**) matte YX.

Semi-transparent surfaces were detected for the specimens printed in glossy mode. The lateral shape of the airfoils was analyzed, especially on the leading edge and the trailing edge of the airfoil (Figures 19 and 20). A rounded edge was detected around the airfoil curve. This is represented by a black border (Figure 19a,b and Figure 20a,b) in the case of glossy type. In addition, the edges of the airfoils printed in matte mode are rounded, as shown in Figure 19c,d and Figure 20c,d.

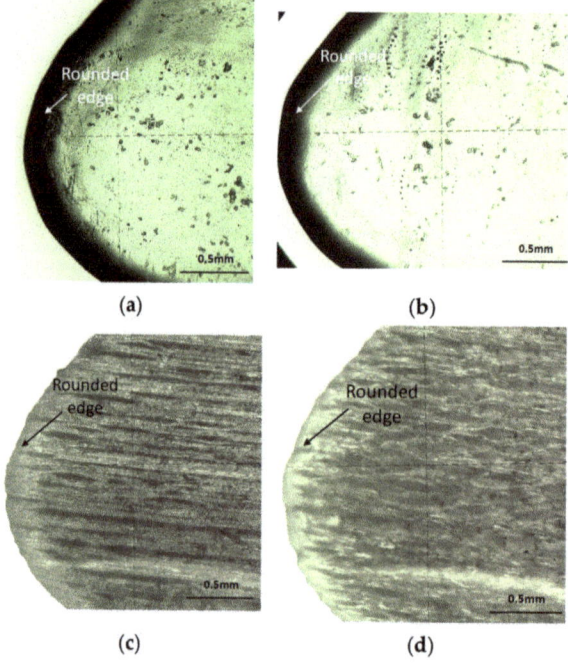

Figure 19. Lateral surface view of the airfoil on the leading-edge zone: (**a**) glossy XY; (**b**) glossy YX; (**c**) matte XY; (**d**) matte YX.

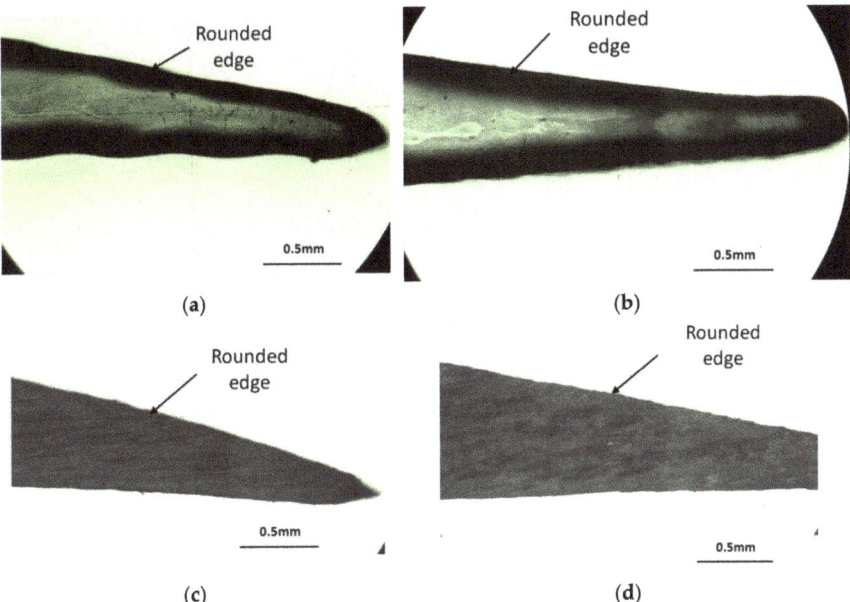

Figure 20. Lateral surface view of the airfoil on trailing edge zone: (**a**) glossy XY; (**b**) glossy YX; (**c**) matte XY; (**d**) matte YX.

The smooth curve of the airfoil was detected for matte-printed specimens (Figure 19c,d and Figure 20c,d). The airfoil curve of glossy specimens presents some deviations from the theoretical profile in the lower part of the trailing edge, as is shown in Figure 20a.

4. Conclusions

There is a need for a universal roughness scale that can describe every type of roughness for different materials used in aerodynamic studies. This paper contributes to the characterization of the surface quality (roughness) of airfoils and wings made by the material jetting AM process. The new methodology based on aerodynamic interest locations and DOE can be used to characterize aerodynamic parts build by AM processes.

The following conclusions can be drawn:

- Material jetting is a quick and simple additive manufacturing process to produce aerodynamic models from polymers.
- The proposed methodology may assess the aerodynamic surface quality in a simple way based on a measurement scheme of roughness on airfoils.
- An inhomogeneous surface roughness distribution on an airfoil was obtained by PolyJet technology using an EDEN 350 system, which can be explained by different surface slopes on the airfoil zone such as the leading edge, central zone, and trailing edge.
- Based on preferential orientations on the build platform, a similar quality of the upper and lower surface of the airfoil was found, both for matte and glossy-printed samples. This could be a beneficial advantage for future aerodynamic studies.
- The experimental roughness (Ra) values of the airfoil printed in PolyJet matte finish were found in the range of 1.06 to 3.62 microns for the YX orientation and 1.74 to 2.46 microns for the XY orientation. The roughness of the airfoils printed in glossy finish presented higher values than matte finish airfoil, in the range of 5.72 to 11.3 microns for the XY orientation and 6.4 to 8.68 microns for the YX orientation.

- The disadvantage of the glossy finish includes some surface quality issues as rough surface areas on the airfoil surface, which were determined by visual inspection, microscopy, and theoretical studies.
- The most influential factor on airfoil surface roughness for the PolyJet process was surface finish type, which was determined from DOE investigation.
- Based on the simulations and the results obtained from the screening DOE, an optimal 3D-printing configuration for airfoils manufactured by PolyJet technology was determined to be XY matte. In addition, the microscopy studies showed that the airfoils printed in matte mode present a homogeneous surface.

Additional study could investigate the dimensional accuracy of the airfoils built by 3D printing, using quality control tools such as the 3D-scanning technique [44]. In addition, future research will investigate other AM processes and materials that can be candidates for airfoils and wings manufacturing based on the proposed methodology.

Funding: This research received no external funding.

Institutional Review Board Statement: Not applicable.

Informed Consent Statement: Not applicable.

Data Availability Statement: Not applicable.

Acknowledgments: The author acknowledges Transilvania University of Brașov for providing the infrastructure used in this work.

Conflicts of Interest: The author declares no conflict of interest.

References

1. Kang, H.S.; Lee, J.Y.; Choi, S.; Kim, H.; Park, J.H.; Son, J.Y.; Kim, B.H.; Noh, S.D. Smart manufacturing: Past research, present findings, and future directions. *Int. J. Precis. Eng. Manuf. Green Technol.* **2016**, *3*, 111–128. [CrossRef]
2. ISO/ASTM 52900-15; Standard Terminology for Additive Manufacturing–General Principles–Terminology. ISO/ASME International: Geneva, Switzerland, 2015.
3. Udroiu, R.; Braga, I.C. System Performance and Process Capability in Additive Manufacturing: Quality Control for Polymer Jetting. *Polymers* **2020**, *12*, 1292. [CrossRef] [PubMed]
4. Tofail, S.A.M.; Koumoulos, E.P.; Bandyopadhyay, A.; Bose, S.; O'Donoghue, L.; Charitidis, C. Additive manufacturing: Scientific and technological challenges, market uptake and opportunities. *Mater. Today* **2018**, *21*, 22–37. [CrossRef]
5. Kadivar, M.; Tormeya, D.; McGranaghan, G. A review on turbulent flow over rough surfaces: Fundamentals and theories. *Int. J.* **2021**, *10*, 100077. [CrossRef]
6. Petrie, H.L.; Deutsch, S.; Brungart, T.A.; Fontaine, A.A. Polymer drag reduction with surface roughness in flat-plate turbulent boundary layer flow. *Exp. Fluids* **2003**, *35*, 8–23. [CrossRef]
7. Salazar, F.; Barrientos, A. Surface Roughness Measurement on a Wing Aircraft by Speckle Correlation. *Sensors* **2013**, *13*, 11772–11781. [CrossRef]
8. Udroiu, R. Applications of additive manufacturing technologies for aerodynamic tests. *Acad. J. Manuf. Eng.* **2010**, *8*, 96–101.
9. Havaldar, S.; Pawar, S.; Lele, A.; Pradhan, R.; Rishi, A. Experimental investigation of lift for NACA 2412 airfoil without internal passage with NACA 2412 airfoil with internal passage in a subsonic wind tunnel. *J. Aerosp. Eng. Technol.* **2015**, *5*, 27–33.
10. Olasek, K.; Wiklak, P. Application of 3D printing technology in aerodynamic study. *J. Phys. Conf. Ser.* **2014**, *530*, 012009. [CrossRef]
11. Lipian, M.; Kulak, M.; Stepien, M. Fast Track Integration of Computational Methods with Experiments in Small Wind Turbine Development. *Energies* **2019**, *12*, 1625. [CrossRef]
12. Junka, S.; Schröder, W.; Schrock, S. Design of additively manufactured wind tunnel models for use with UAVs. *Procedia CIRP* **2017**, *60*, 241–246. [CrossRef]
13. Kim, G.D.; Oh, Y.T. A benchmark study on rapid prototyping processes and machines: Quantitative comparisons of mechanical properties, accuracy, roughness, speed, and material cost. *Proc. Imeche Part B J. Eng. Manuf.* **2008**, *222*, 201–215. [CrossRef]
14. Mueller, J.; Shea, K.; Daraio, C. Mechanical properties of parts fabricated with inkjet 3d printing through efficient experimental design. *Mater. Des.* **2015**, *86*, 902–912. [CrossRef]
15. Zaharia, S.M.; Enescu, L.A.; Pop, M.A. Mechanical Performances of Lightweight Sandwich Structures Produced by Material Extrusion-Based Additive Manufacturing. *Polymers* **2020**, *12*, 1740. [CrossRef]
16. Leach, R. Metrology for additive manufacturing. *Meas. Control* **2016**, *49*, 132–135. [CrossRef]
17. Krolczyk, G.; Raos, P.; Legutko, S. Experimental analysis of surface roughness and surface texture of machined and fused deposition modelled parts. *Tehnički Vjesnik* **2014**, *21*, 217–221.

18. Umaras, E.; Tsuzuki, M.S.G. Additive manufacturing-considerations on geometric accuracy and factors of influence. *IFAC-PapersOnLine* **2017**, *50*, 14940–14945. [CrossRef]
19. Moylan, S. Progress toward standardized additive manufacturing test artifacts. In Proceedings of the ASPE 2015 Spring Topical Meeting Achieving Precision Tolerances in Additive Manufacturing, Raleigh, NC, USA, 26–29 April 2015; pp. 100–105.
20. Yap, Y.L.; Wang, C.; Sing, S.L.; Dikshit, V.; Yeong, W.Y.; Wei, J. Material jetting additive manufacturing: An experimental study using designed metrological benchmarks. *Precis. Eng.* **2017**, *50*, 275–285. [CrossRef]
21. Udroiu, R.; Braga, I.C.; Nedelcu, A. Evaluating the Quality Surface Performance of Additive Manufacturing Systems: Methodology and a Material Jetting Case Study. *Materials* **2019**, *12*, 995. [CrossRef] [PubMed]
22. ISO/ASTM 52902-15; Additive Manufacturing–Test Artifacts–Geometric Capability Assessment of Additive Manufacturing Systems. ISO/ASME International: Geneva, Switzerland, 2019.
23. Perez, M.; Medina-Sánchez, G.; García-Collado, A.; Gupta, M.; Carou, D. Surface quality enhancement of fused deposition modeling (FDM) printed samples based on the selection of critical printing parameters. *Materials* **2018**, *11*, 1382–1395. [CrossRef]
24. Strano, G.; Hao, L.; Everson, R.M.; Evans, K.E. Surface roughness analysis, modelling and prediction in selective laser melting. *J. Mater. Process. Technol.* **2013**, *213*, 589–597. [CrossRef]
25. Chen, Y.; Lu, J. RP Part Surface quality versus build orientation: When the layers are getting thinner. *Int. J. Adv. Manuf. Technol.* **2013**, *67*, 377–385. [CrossRef]
26. Nagendra, K.M.; Rastogi, V.; Singh, P. Comparative Study and Measurement of Form Errors for the Component Printed by FDM and PolyJet Process. *Instrum. Mes. Métrologie.* **2019**, *18*, 353–359. [CrossRef]
27. Kumar, K.; Kumar, G.S. An experimental and theoretical investigation of surface roughness of poly-jet printed parts. *Virtual Phys. Prototyp.* **2015**, *10*, 23–34. [CrossRef]
28. Kechagias, J.; Stavropoulos, P. An investigation of sloped surface roughness of direct poly-jet 3D printing. In Proceedings of the International Conference on Industrial Engineering—INDE 2015, Zakynthos, Greece, 16–20 July 2015; pp. 150–153.
29. Canellidis, V.; Giannatsis, J.; Dedoussis, V. Genetic-algorithm based multi-objective optimization of the build orientation in stereolithography. *Int. J. Adv. Manuf. Technol.* **2009**, *45*, 714–730. [CrossRef]
30. Gülcan, O.; Günaydın, K.; Tamer, A. The State of the Art of Material Jetting—A Critical Review. *Polymers* **2021**, *13*, 2829. [CrossRef]
31. Turek, P.; Budzik, G.; Sęp, J.; Oleksy, M.; Józwik, J.; Przeszłowski, Ł.; Paszkiewicz, A.; Kochmański, Ł.; Żelechowski, D. An Analysis of the Casting Polymer Mold Wear Manufactured Using PolyJet Method Based on the Measurement of the Surface Topography. *Polymers* **2020**, *12*, 3029. [CrossRef] [PubMed]
32. Montgomery, D.C. *Design and Analysis of Experiments*; John Wiley & Sons: Hoboken, NY, USA, 2017; ISBN 9781119113478.
33. ISO/ASTM 52921-13; Standard Terminology for Additive Manufacturing—Coordinate Systems and Test Methodologies. ISO: Geneva, Switzerland, 2013.
34. Minitab. Getting Started with Minitab 17. Available online: https://www.minitab.com (accessed on 10 September 2018).
35. Objet Geometries. Eden 500V/350V/350 3-D Printer System. In *User Guide*; Objet Geometries Ltd.: Rehovot, Israel, 2007.
36. Stratasys. PolyJet Materials Data Sheet. Available online: http://www.stratasys.com (accessed on 10 September 2015).
37. Kepler, J. Investigation of Acrilic Based Systems for 3D Polyjet Printing. Master's Thesis, Degree-Granting University of Linz, Linz, Austria, 2018.
38. Lee, J.M.; Zhang, M.; Yeong, W.Y. Characterization and evaluation of 3D printed microfluidic chip for cell processing. *Microfluid. Nanofluid* **2016**, *20*, 5. [CrossRef]
39. Peng, T.; Leary, P.; Tran, P. PolyJet 3D Printing of Composite Materials: Experimental and Modeling Approach. *JOM Min. Met. Mater. Soc.* **2020**, *72*, 1105–1117. [CrossRef]
40. Aly, K.I.; Abd El-Lateef, H.M.; Yehia, N.; Khodairy, A.; Sayed, M.M.; Ali, M.A.E.A.A. Novel polyesters based on indazole moiety: Synthesis, characterization and applicability as efficient inhibitors for acidic X-65-steel corrosion. *React. Funct. Polym.* **2021**, *166*, 105001. [CrossRef]
41. Derby, B. The inkjet printing of functional and structural materials: Fluid property requirements, feature stability, and resolution. *Annu. Rev. Mater. Res.* **2010**, *40*, 395–414. [CrossRef]
42. ISO 4288:1996; Geometrical Product Specifications (GPS)—Surface Texture: Profile Method—Rules and Procedures for the Assessment of Surface Texture. ISO/ASME International: Geneva, Switzerland, 1996.
43. Automotive Industry Action Group. *Measurement Systems Analysis—Reference Manual*, 4th ed.; MSA-4; Automotive Industry Action Group (AIAG): Southfield, MI, USA, 2010.
44. Bere, P.; Neamtu, C.; Udroiu, R. Novel Method for the Manufacture of Complex CFRP Parts Using FDM-based Molds. *Polymers* **2020**, *12*, 2220. [CrossRef] [PubMed]

Article

Mechanical Properties and In Vitro Evaluation of Thermoplastic Polyurethane and Polylactic Acid Blend for Fabrication of 3D Filaments for Tracheal Tissue Engineering

Asmak Abdul Samat [1,2], Zuratul Ain Abdul Hamid [3], Mariatti Jaafar [3] and Badrul Hisham Yahaya [1,*]

1. Lung Stem Cell and Gene Therapy Group, Regenerative Medicine Cluster, Advanced Medical and Dental Institute (IPPT), Sains@Bertam, Universiti Sains Malaysia, Kepala Batas 13200, Malaysia; asmakas@iium.edu.my
2. Fundamental Dental and Medical Sciences, Kulliyyah of Dentistry, International Islamic University Malaysia, Kuantan 25200, Malaysia
3. School of Materials and Mineral Resources Engineering, Universiti Sains Malaysia, Nibong Tebal 14300, Malaysia; srzuratulain@usm.my (Z.A.A.H.); mariatti@usm.my (M.J.)
* Correspondence: badrul@usm.my

Citation: Abdul Samat, A.; Abdul Hamid, Z.A.; Jaafar, M.; Yahaya, B.H. Mechanical Properties and In Vitro Evaluation of Thermoplastic Polyurethane and Polylactic Acid Blend for Fabrication of 3D Filaments for Tracheal Tissue Engineering. *Polymers* **2021**, *13*, 3087. https://doi.org/10.3390/polym13183087

Academic Editors: Swee Leong Sing and Wai Yee Yeong

Received: 15 August 2021
Accepted: 7 September 2021
Published: 13 September 2021

Publisher's Note: MDPI stays neutral with regard to jurisdictional claims in published maps and institutional affiliations.

Copyright: © 2021 by the authors. Licensee MDPI, Basel, Switzerland. This article is an open access article distributed under the terms and conditions of the Creative Commons Attribution (CC BY) license (https://creativecommons.org/licenses/by/4.0/).

Abstract: Surgical reconstruction of extensive tracheal lesions is challenging. It requires a mechanically stable, biocompatible, and nontoxic material that gradually degrades. One of the possible solutions for overcoming the limitations of tracheal transplantation is a three-dimensional (3D) printed tracheal scaffold made of polymers. Polymer blending is one of the methods used to produce material for a trachea scaffold with tailored characteristics. The purpose of this study is to evaluate the mechanical and in vitro properties of a thermoplastic polyurethane (TPU) and polylactic acid (PLA) blend as a potential material for 3D printed tracheal scaffolds. Both materials were melt-blended using a single screw extruder. The morphologies (as well as the mechanical and thermal characteristics) were determined via scanning electron microscopy (SEM), Fourier Transform Infrared (FTIR) spectroscopy, tensile test, and Differential Scanning calorimetry (DSC). The samples were also evaluated for their water absorption, in vitro biodegradability, and biocompatibility. It is demonstrated that, despite being not miscible, TPU and PLA are biocompatible, and their promising properties are suitable for future applications in tracheal tissue engineering.

Keywords: thermoplastic polyurethane; polylactic acid; trachea scaffold; 3D filament

1. Introduction

Tracheal injury can result from several conditions, including cancer, infection, trauma, or congenital anomalies. The conventional indication for therapy in severely injured tracheas of any aetiology is partial or full reconstruction, which necessitates the substitution of a graft or scaffold at the site of the lesion [1,2]. Unfortunately, even though there are a few treatment options available such as natural grafts or synthetic replacement, no optimal material has met the criteria. The limitations for natural grafts include the availability of donors and the unmatched size of donor grafts. According to the Organ Procurement & Transplantation Network, United States Department of Health and Services, the number of patients on the national transplant waiting list until July 2019 for all organ types has increased to more than 100,000. Out of this number, two-thirds are above the age of 50, while almost 2000 are below the age of eighteen and only one-third of the total numbers received organ transplantation [3]. In addition, the natural grafts derived from donors are challenged by the possibility of severe immune-rejection risks and complications caused by infection or disease from the donor-to-patient [4]. On the other hand, synthetic scaffolds are commonly associated with the biocompatibility of the scaffold material, inadequate mechanical properties, and biodegradability over time. In addition, other problems (such as availability for mass production with easy fabrication and the need for the size of the trachea should be custom-made to patients) persist [5–7].

Synthetic materials such as biodegradable polymers are gaining attention as materials in tissue engineering due to their broad processability window where the macro- and microstructures, mechanical properties, and degradation time can be easily manipulated and controlled. Scaffolds are fabricated and manipulated through various techniques to produce high precision, which suits the application. Some of the techniques used in the fabrication of tracheal scaffold using biodegradable materials that have been tested in animal models are electrospinning [8], thermally induced phase separation [9], and three dimensional (3D) printed technology (additive manufacturing) [10,11]. However, the choice of material and design for tracheal scaffold fabrication remains a challenge. To meet the selection criteria, many requirements must be fulfilled. For example, the scaffolds must create a suitable 3D niche for the cells to grow, proliferate, and differentiate, and should not elicit an immune reaction that can trigger a severe inflammatory response that might reduce healing or cause rejection by the body [12]. In vivo, the scaffold functions as a temporary framework that degrades over time which is eventually replaced by the body's cells. Therefore, the degradation rate should match the rate at which the cells produce their cellular matrix, while the by-products released should be innocuous and eliminated safely through the body system [13]. Similarly, sufficient mechanical integrity of the implanted scaffold is required to allow for physiological functionality starting from implantation until the completion of the remodelling process.

Additive manufacturing (AM), also known as three-dimensional (3D) printing, has been used to fabricate tissue-engineered constructs. According to the ISO/ASTM standard, AM is defined as the "process of joining materials to make parts from 3D model data, usually layer upon layer" [14]. The AM is significantly different from traditional formative or subtractive manufacturing. It is the closest to 'bottom up' manufacturing, in which a structure can be built into its intended shape using a layer-by-layer technique. This layer-by-layer manufacturing technique enables unprecedented precision and control for constructing complex, composite, and hybrid structures. The four key components in AM include a digital model of the object, materials that are consolidated from the smallest possible form, a machine for laying materials, and a digital control system for the machine to lay the materials layer-by-layer to form a complex structure with customizable shape, size, and internal architecture [15–17]. The 3D fabrication of a tracheal scaffold has been reported in several preclinical studies using different types of polymeric materials such as polylactic acid [18] and polycaprolactone [19,20].

Thermoplastic polyurethane (TPU) is a polymeric material that can be manipulated, moulded, and produced through heating in various industrial processes. Polyurethane is composed of three materials; a diisocyanate, a chain extender and a macrodiol (or polyol) which are linked to form linear, segmented copolymers consisting of alternating hard and soft segments. The soft and flexible segment is derived from polyols such as polyester, while the rigid and hard segment is formed from the diisocyanate and chain extender [21]. TPU exhibits a broad range of mechanical properties across a wide range of temperatures due to the various ratios of soft to hard segments. As a result of its excellent physical properties and biocompatibility, it is widely used in biomedical applications, particularly in flexible uses such as blood vessels [22–24], catheters [25,26], and cartilage [27,28].

Polylactic acid (PLA) is a semi-crystalline polymer that belongs to the α-hydroxy acid family, derived from renewable sources such as corn, potatoes, sugarcane, and beets. It is classified as an aliphatic polyester because of the ester bonds that connect the monomer units, the lactic acids [29,30]. PLA and its copolymers have become one of the most attentively studied components in the biomedical field because of their excellent biological and mechanical properties, biodegradability and processability. Hence, it has wide applications such as medical implants, sutures [31], bone fixation screws [32], and drug delivery systems [33]. However, biodegradable PLA exhibits little to no elastic behaviour and is not favoured for applications requiring high flexibility or deformation in situ. Furthermore, the inherent hydrophobicity and slow degradability of PLA slightly impede its application in biological systems [34].

Blending two or more polymers is a common physical modification approach to enhance the existing properties of both materials to customize the desired properties for a particular application [34–36]. The blending technique has been utilized to overcome the limitations of the physical properties of polymers and has resulted in materials with novel properties such as shape memory and morphology that are not present in the parent polymers. These materials can be moulded into various structures, including films, porous scaffolds, fibres, filaments, and particles, depending on the intended application, with properties tuned for use in a variety of biomedical applications. Polymer blending facilitates the efficient and cost-effective modification or improvement of a polymer's properties, thereby minimizing the significant costs and efforts associated with research and development of new polymers or copolymers. The blending techniques used are melt extrusion, foaming, electrospinning, and compression moulding [34,35].

Several biodegradable polymers that have been used to fabricate a tracheal scaffold are polylactic acid (PLA), polyglycolic acid (PGA), polylactide-co-glycolide acid (PLGA) [37–40], polypropylene [41,42], polyethylene terephthalate [43], high-density polyethylene (HDPE) [44], and polycaprolactone (PCL) [45–48]. Most of these polymers are used in combination with other synthetic or natural materials to enhance their properties.

Despite a large amount of research into various biodegradable polymers, clinical performance has yet to satisfy theoretical expectations. As a result, there is currently no clinically feasible solution for patients with long segmental airway problems. Therefore, an ideal synthetic scaffold that is biocompatible, timely degraded, and eliminated by the body system with appropriate and physical-mechanical qualities that can be easily replicated when needed (and is maybe individually custom-made to prevent prosthesis failure) is required. This study investigated the physical blending of two materials to obtain the optimum mechanical properties while retaining the material's superior inherent properties. Additionally, it aimed to evaluate the physical and mechanical properties of a series of TPU and PLA blends, which will be used to produce filaments for 3D printing for tracheal tissue engineering. The TPU/PLA blended matrix, a combination of soft material TPU and rigid material PLA, is expected to act as an artificial ECM by possessing suitable mechanical strength and flexibility between the TPU and PLA.

2. Experimental

2.1. Materials

TPU Estane 58,311 NAT 028 (Brussel, Belgium); PLA NatureWorks, 2002D was purchased from NatureWorks LLC (Minnesota, MN, USA) with a specific gravity of 1.24 and a melt index of 5.0–7.0 g/10 min (2.16 kg loads at 210 °C).

2.2. Methods

2.2.1. Fabrication of Polymer Blends' Filaments via Melt Extrusion Technique

Prior to extrusion, the pellets of both polymers were dried in a 60 °C oven for 12 h. Then, the extrusion of fibres was performed using a Brabender (Duisburg, Germany) single screw extruder with a 1.75 mm die, operated according to the manufacturer's instructions. A total weight of 100 g was used for each composition based on their weight percentage ratio (TPU: PLA) and coded as 100/0, 90/10, 80/20, 70/30, 60/40 and 0/100, respectively. Next, both materials were manually premixed via tumbling in a plastic zip-lock bag before melt-compounding. Once optimised, the temperature of the single screw extruder was set at 170° to 205 °C ($\pm$5 °C), the rotation speed was at 40 ($\pm$5) rpm, and the mixture was fed for melt compounding. Finally, the filaments were pelletised and hot-pressed into dumbbell shapes and 10 mm × 10 mm square samples, allowing for various characterisation methods. The TPU filament was produced using TPU pellets only in the same manner as other blends.

2.2.2. Characterisation of Polymer Blends

Fourier Transform Infrared (FTIR)

FTIR spectra were obtained in a reflective absorbance mode on a Perkin Elmer spectrometer (Waltham, MA, USA) with a constant spectral resolution of 4 cm^{-1}, in the range of 4000 to 550 cm^{-1}, and after 16 scans. The reported spectra were analysed quantitatively using Perkin Elmer Software (Waltham, MA, USA) version 10.

Differential Scanning Calorimetry (DSC)

Thermal property analyses were carried out using a Differential Scanning Calorimetry (Mettler Toledo, Greifensee, Switzerland). In standard aluminium pans, about 10 mg of the samples were from room temperature to 250 °C at a rate of 20 °C/min and held isothermally for 5 min to exclude all previous thermal history. After cooling at 5 °C/min to −80 °C at a rate of 10 °C/min, samples were heated again at 20 °C/min to 250 °C. All of the experiments were conducted under a nitrogen atmosphere.

Mechanical Testing

The mechanical properties of all blends were conducted in tensile uniaxial mode using an Instron universal testing machine model 3366 (Norwood, MA, USA) at a crosshead speed of 5 mm/min according to ASTM D638. The samples analysed were 60 mm × 5 mm in size, with a thickness of around 1.0 mm. The slope of the straight-line stress-strain curve was used to calculate the Young's Modulus (YM) of the polymer blends and the effects of tensile strength and percentage of elongation at break. The mean and standard deviation of five measures were used to calculate all results.

Scanning Electron Microscope (SEM)

SEM images were captured using a tabletop SEM (Hitachi, Tokyo, Japan). The cross-sectional surfaces were obtained from a tensile examination of a broken dumbbell. Before microscopy experiment, all specimens were sputtered with a thin layer of gold.

Water Absorption Study

The water absorption test was performed on 10 mm square samples according to ASTM D570. The samples were dried at 60 °C for 24 h to achieve a stable weight. The dried samples were weighed and placed in the test plate wells, prewetted with PBS solution before filling with 5 mL of PBS, pH 7.4 at 37 °C. The samples were incubated and tested at 8, 24, 48 and 72 h. The excess water was carefully removed with tissue paper, and the samples were re-weighed. Water absorption was calculated based on the amount of water absorbed according to Equation (1):

$$Water\ absorption\ (\%) = \frac{Wwet - Wdry}{Wdry} \times 100 \qquad (1)$$

For each composition, three specimens were examined to achieve an average value.

In Vitro Degradation Study

PBS soaking tests were used to mimic the hydrolytic degradation activity of TPU/PLA blends and control scaffolds. The samples were weighed after drying overnight at 60 °C. Each sample was individually enclosed in a plastic container filled with a 1X PBS solution and incubated at 37 °C in an orbital shaker (Stuart S1500, Illinois City, IL, USA) at a shaking rate of 50 rpm. Every 7 days, PBS was refreshed, and the test lasted up to 7 months. The samples were rinsed three times with purified water before being dried overnight and weighed at each time point. Equation (2) was used to measure the weight loss of the materials.

$$Degradation\ (\%) = \left(1 - \frac{W_n}{W_o}\right) \times 100 \qquad (2)$$

W_0 is the initial sample weight, and W_n is the weight of the same sample after degradation for a time, n.

In Vitro Biocompatibility Assay

The blended samples were evaluated for biocompatibility and possible use as tissue engineering scaffolds in biomedical applications.

Pellet Sterilisation

Prior to in vitro testing, the TPU/PLA pellets were sterilised by being immersed in 70% ethanol (v/v) for 2 h, followed by rinsing three times with 1x PBS to eliminate all traces of ethanol. The pellets were then air-dried in a sterile atmosphere before being sterilised for 2 h with ultraviolet light. This phase ensured that any pollutants on the surface of the pellets were removed.

Pesto Blue Viability Assay

In order to be considered for biomedical applications, any material must not impose any toxicity to the surrounding tissues. The toxicity of the TPU and PLA was investigated in this study and was defined as a reduction in cell growth to less than 50% viability. The sterile pellets were immersed in complete α-MEM overnight before testing. BEAS-2-B (human bronchial epithelial cells) were purchased form American Type Culture Collection (ATCC, Manassas, VA, USA) were grown to confluence in complete α-MEM containing 10% FBS and 1% antibiotic–antimycotic (AA) solution at 37 °C in an incubator with 5% carbon dioxide (CO_2). In 24-well plates, one pellet was placed in each well, followed by direct cell seeding on top of the pellet at a seeding number of 1×10^4 cells per well, and the plates were incubated in a CO_2 incubator for 3 days. As a positive control, a tissue culture plate with fresh α-MEM was used. The toxicity test was performed using Presto Blue Cell Viability Reagent on days 1, 2 and 3. The metabolic product of viable cells was released into the culture medium, reduced resazurin to resorufin, and changed the colour from blue to pinkish red. When reduction did not occur in a nonviable setting, the blue-coloured resazurin was preserved. An automatic ELISA reader was used to test the fluorescence of the samples in triplicate. Equation (3) was used to measure the cell toxicity percentage:

$$\text{Cell toxicity (\%)} = \frac{Fluorescence_{treatment} - Fluorescence_{control}}{Fluorescence_{control}} \times 100 \quad (3)$$

Statistical Analysis

The mean and standard deviation of all the results were calculated (SD). The one-way analysis of variance (ANOVA) was used for normally distributed data, while for non-normally distributed data, the nonparametric test was used. The Tukey's test was then used to assess the data's particular differences, with $p < 0.05$ indicating statistical significance. The data were analysed using GraphPad Software (Prism 9.0, GraphPad Software, La Jolla, CA, USA).

3. Results and Discussion

Apart from being biocompatible and biodegradable, as the foundation of the tracheal structure, the material of the scaffold should have adequate mechanical strength and flexibility to enable physiological function during breathing. The rigid components of the cartilage retain the trachea lumen open and prevent its collapse under negative air pressure, preventing airflow limitations [49]. The purpose of this study is to determine the physical and mechanical properties, as well as the absorption, in vitro degradation, and biocompatibility, of a series of TPU and PLA blends that will be produced as filaments and 3D printed as a potential material for tracheal replacement.

3.1. Fabrication of Filament and Identification of the Materials

The physical blending of the material through the melt extrusion technique of TPU and PLA was chosen in this study to produce filaments that will be fabricated as tracheal scaffold using a 3D printing technique. Both materials were mixed based on their weight percentage and characterised accordingly. FTIR transmission spectra of the samples are presented in Figure 1 with main characteristic bands of pure TPU appeared at 3328 cm^{-1}, 2935 cm^{-1}, and 2850 cm^{-1} which correspond to stretching of -NH- in urethane and asymmetric and symmetric vibrations in -CH2- respectively [50–53]. In addition, 1700, 1531 and 1314 cm^{-1} bands were associated with bending and stretching of amide I, II and III bonds. The intensity of characteristic bands in TPU reduces as the PLA contents increases and vice versa. The characteristic bands of PLA were seen at 1750 cm^{-1}, 1456 cm^{-1} and 1180 cm^{-1}, corresponding to asymmetric vibration of -C=O and the asymmetric and symmetric stretching of -C-O-C bonds, respectively. The characteristic bands and their activities are summarised in Table 1. Overall, the spectra revealed no new chemical bonds, suggesting that both PLA and TPU were successfully compounded during melt blending [54,55]. Physical blending does not alter the chemical characteristic of the constituents of the polymers [34]. Hence, the properties of the blend can be conveniently customized using different compositions of the polymer to suit its application.

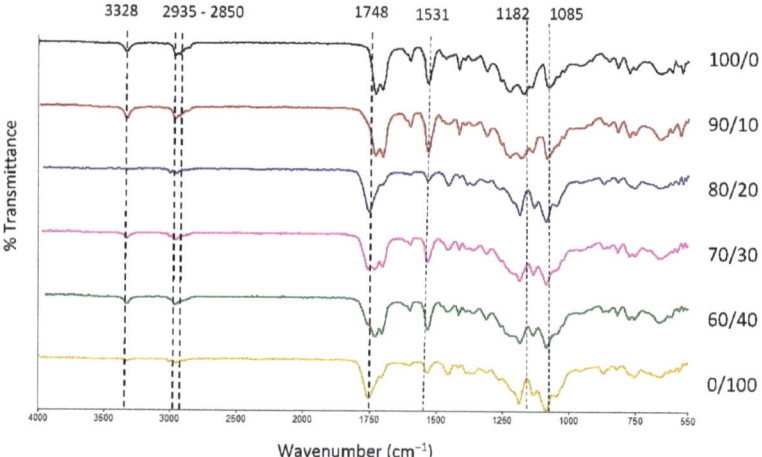

Figure 1. FTIR spectra of TPU, PLA, and TPU/PLA blends.

Table 1. Characteristic bands and corresponding activities of TPU and PLA.

Characteristic Band	Activity	Material
3328	Stretching -NH- in urethane	TPU
2935	Asymmetric vibration in -CH2-	TPU
2850	Symmetric vibration in -CH2-	TPU
1748	-C=O stretching (amide I)	TPU/PLA
1531	N-H bending vibration (amide II)	TPU
1314	C-N (amide III)	TPU
1182	asymmetric stretching of -C-O-C-	PLA
1085	symmetric stretching of -C-O-C-	PLA

3.2. Morphology and Miscibility of the Blends

The SEM images of the fractured surface TPU/PLA dumbbell films were analysed to study the phase morphology of the samples. In contrast, the DSC findings were used to determine the thermal properties and miscibility of the samples. As shown in Figure 2, the fractured surface morphology of pure TPU and pure PLA reveals the homogeneous distribution of fibrous TPU and continuous matrix of PLA, respectively. In contrast, in blended compositions, the smooth-edged PLA domains are distributed in the fibrous TPU matrix, resulting in a notably heterogeneous two-phase structure. Both 90/10 and 80/20 samples displayed more fibrous TPU than 70/30 and 60/40 samples, reflecting the flexibility of the blends. However, several separated phase domains with minute debonded holes were also observed, indicating a weak interfacial contact between the TPU matrix and the PLA domain. As the PLA ratio increased, larger and more PLA spheres were detected in the TPU matrix, although they were dispersed equally. All blended polymers exhibit phase separation, suggesting that TPU and PLA are immiscible.

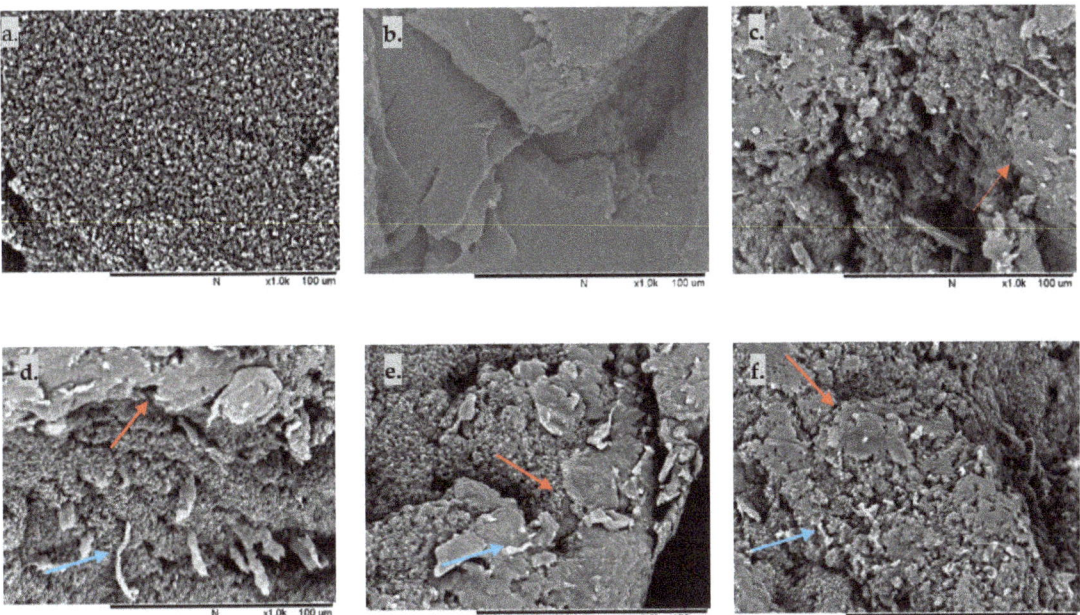

Figure 2. Cross-section SEM images of reactively extruded films: (**a**) 100/0, (**b**) 0/100. Homogeneous matrices are present in pure TPU and pure PLA (**c**) 90/10, (**d**) 80/20, (**e**) 70/30, (**f**) 60/40. Some fibrous TPU is present in the polymer blends, with PLA domains are dispersed in TPU matrices in all blends. Red arrows show PLA particles in the TPU matrix, while blue arrows indicate the fibrous TPU of the fractured surface. Scale bar of 100 μm.

TPU is composed of diisocyanate hard segments and polyester macroglycol soft segments, while PLA is an aliphatic polyester. Due to their structures, two glass transition temperature (T_g) and melting temperature (T_m) values were observed in pure TPU, which correspond to hard and soft segments as shown in DSC curves in Figure 3. On the other hand, only one T_g and T_m were seen in pure PLA.

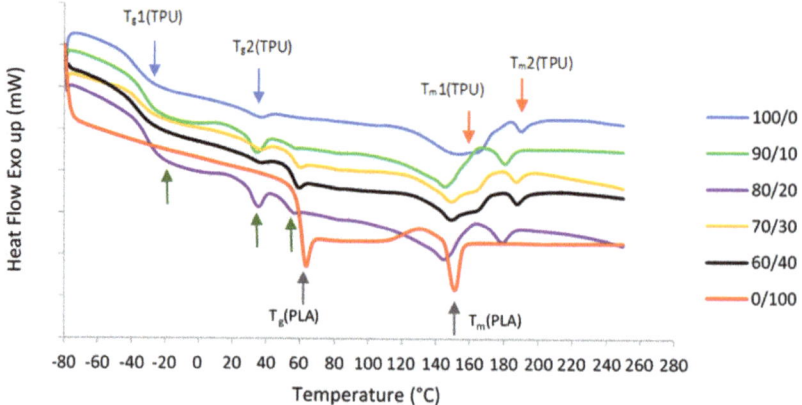

Figure 3. DSC secondary melting curves of pure TPU and PLA, and TPU/PLA blends. The blue arrows show the first and second T_g values of pure TPU, while red arrows indicate both T_m values of pure TPU. Grey arrows show the T_g and T_m values of pure PLA. Finally, the green arrows depict all three T_g values of TPU/PLA blends.

The immiscibility of the blends is further demonstrated by DSC curves and summarized data as shown in Table 2. The T_g values can determine the extent of blend miscibility, partial miscibility, or total immiscibility as a function of the polymer blend composition. Normal T_g values of the TPU and PLA ranges from around −18 °C to −47 °C and 61 °C to 67 °C, respectively [56,57]. Miscible polymers typically have a single T_g, whereas immiscible blends transition temperatures shift toward each other to a degree depending on the mutual miscibility of the phases [58–61]. In the event of a completely immiscible blend system, the blend components may remain their original T_g values, regardless of the blend composition [34]. In this study, the appearance of three distinct T_g values in 90/10, 80/20, 70/30, and 60/40 samples suggest that TPU and PLA are immiscible. Two T_g values correspond to the TPU structure, whereas another T_g value comes from the PLA. The immiscibility was demonstrated by considerable macroscopic phase separation in the blends in SEM. Moreover, all mixed samples that confirm the immiscibility also show two different melting temperatures (T_m). This result is in line with [51,61–63]. In immiscible blends, the properties of the component polymers integrate in such a way that the blend morphology is a direct representation of the component morphology. As a result, blend morphology is regarded as a good indicator of blend miscibility.

Table 2. Differential scanning calorimetry data for TPU, PLA, and blends.

Sample	$T_{g\,TPU}$	$T_{g\,PLA}$	T_m1	T_m2
100/0	−26.98	-	153.88	189.90
0/100	-	63.16	150.74	-
90/10	−25.12	58.21	145.18	180.20
80/20	−19.70	55.87	144.11	178.44
70/30	−27.62	59.24	148.17	186.84
60/40	−28.00	58.56	148.15	187.15

T_g = glass transition temperature, T_m = melting temperature.

3.3. Mechanical Properties

It is well recognised that the morphology of the materials greatly influences the mechanical properties of blends, and the properties of the final product can be achieved by adjusting the morphology. The tensile strength and YM measurement were derived from the stress–strain curve of the tensile test ASTM D638. Figure 4 depicts the mechanical properties of the TPU/PLA blends. Pure PLA has the highest tensile strength and YM at 46.48 ± 5.51 MPa and 2282.80 ± 95.60 MPa, preceded by TPU blends with higher PLA contents. This means that the higher the PLA concentration, the stiffer the blend. On the other hand, pure PLA demonstrated the least amount of flexibility because it cannot elongate by more than 5% due to its inelastic nature. In comparison, pure TPU showed the highest flexibility (up to more than 25%) without fracture, as shown in Figure 4b. The flexibility decreases with decreasing TPU concentration, with the 60/40 blend exhibiting the least flexibility of all blends. The morphology of TPU and PLA blends appeared to be related to their mechanical properties, with stretched fibrous TPU can be seen in fractured surfaces images of the 90/10 and 80/20 samples reflecting the blends' flexibility.

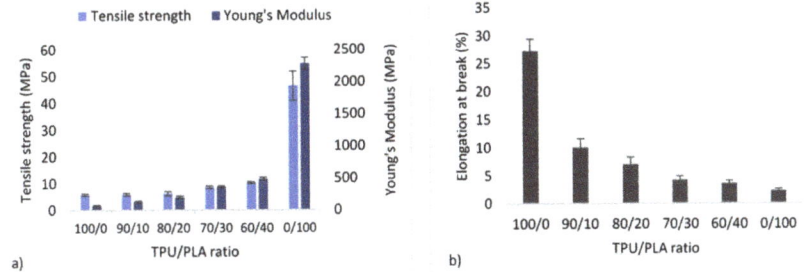

Figure 4. Mechanical properties of TPU/PLA blends showing (**a**) tensile strength and Young's modulus in MPA, and (**b**) percentage of elongation at break.

As a substitute for the cartilage wall, the scaffold is expected to possess adequate strength to keep the airway open and resist collapse. At the same time, it maintains the flexibility to allow flexion/bending despite intrathoracic pressure differences during breathing cycles [63,64]. TPU is known to have flexibility and is widely used in soft tissue engineering [23,24,65,66]. The present study revealed that the tensile strength and Youngs' modulus of TPU were proportionately increased when PLA was added. In contrast, the percent elongation reduces accordingly. The results obtained were similar to those reported by Lis–Bartos et al. (2018) [67] and Mi et al., (2013) [51]. The overall physicomechanical behaviour of immiscible systems is critically dependent on two demanding structural parameters. First, an appropriate interfacial tension leads to a phase size that is small enough to allow the material to be considered macroscopically homogeneous. Secondly, an interphase adhesion is strong enough to assimilate stresses and strains without disrupting the established morphology [68]. Even though the blend was not miscible, it is noted that they considered having good compatibility with each other. This is due to the composition of diisocyanate hard segments and polyester macroglycol soft segments of the TPU and aliphatic polyester in PLA, which forms hydrogen bonding between the molecules of the blend, as shown in Figure 5.

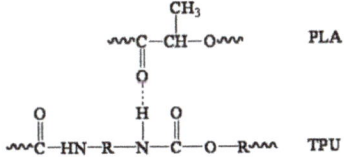

Figure 5. Schematic diagram of hydrogen bonding between the molecules of PLA and TPU. Reproduced with permission from Feng and Ye, Journal of Applied Polymer Science; reprinted with permission from ref [69], Copyright 2011 John Wiley and Sons.

The blend is suitable for the fabrication of a tissue-engineered scaffold since the mechanical properties are within the normal human tracheal range: tensile stiffness 1–15 MPa [70] and YM 12.2–20.5 MPa in young and old humans; ~16 MPa in both human male and female [64].

The purpose of combining two polymers is to produce a material with improved physical properties over the parent polymers. The degree of modification is relative to the quantity of each composition of the polymer. The morphology of the blend and the dispersion of the phases have a considerable effect on the thermomechanical characteristics. For example, TPU with a lower tensile strength is combined with PLA to increase the tensile strength. Despite the modest reduction in elongation at break, the final material can be adjusted by varying the polymer composition to attain the optimal combination of thermal and mechanical properties for specific applications.

3.4. Water Absorption and In Vitro Degradation Rate

An ideal implant material should possess optimal absorption and biodegradable properties similar to the regenerative process of the native tissue [71]. This fills up surrounding tissues and provides a biocompatible framework that allows cells, blood vessels, and newly created tissues to develop and maintain an extracellular matrix [72].

The swelling behaviour of a material (or its ability to absorb water) is an important factor to consider when constructing a scaffold. Although excessive water absorption damages the scaffold's morphology, its absence causes inadequate absorption, inhibiting cell growth in vivo [54]. The absorption rate was determined by immersing the samples in PBS solution. Pure PLA showed the highest absorption rate in PBS solution, which was approximately 5% in the first hour, decreased significantly after 8 h, and remained stable until 48 h. On the other hand, other compositions exhibited nearly equal absorption, varying from 1 to 2 percent, but decreased steadily and remained nearly stable until 48 h. No significant differences were found in any of the compositions. The material's absorption property indicates the scaffold's ability to bind to the surrounding metabolites and promotes the transportation of nutrients and cell integration throughout the scaffold [73].

One of the primary goals of biodegradable tissue-engineered scaffolds is to provide a mechanical structure or framework to support the extracellular matrix (ECM). At the same time, it degrades evenly/slowly via hydrolytic degradation, allowing the surrounding tissue to recover the supporting function of the scaffold [74]. The monomeric components of the polymers are innocuously eliminated through the body system [75]. In the degradation study over 7 months, all of the samples showed a gradual loss with pure TPU (100/0) exhibiting the fastest degradation rate, followed by 90/10, 80/20, 70/30, 60/40 and pure PLA (0/100). The higher the TPU concentration, the faster rate of degradation. The degradation percentage up to 7 months was less than 6%, suggesting that the blended material is long-term stable. In contrast, the degradation of PLA is slower, which started only after 1 month of incubation and gradually degraded at a very slow rate. Up until 7 months, the degradation percentage was approximately 2% and was the lowest among other blends. Figure 6 shows both the rate of absorption and degradation results.

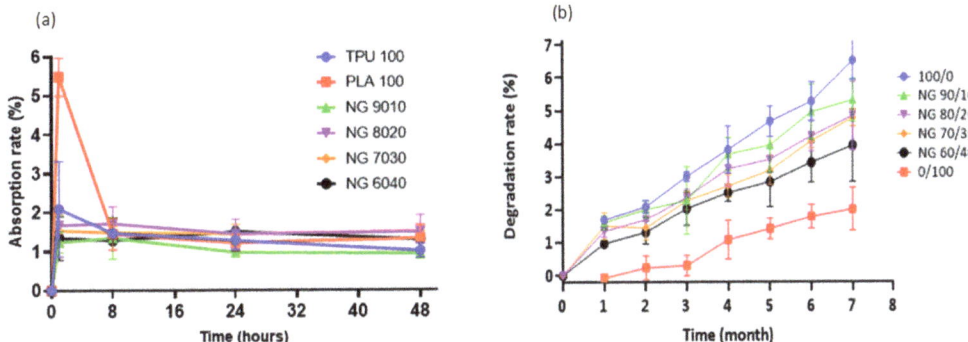

Figure 6. (**a**) Water absorption rate of TPU, PLA, and the blends, (**b**) In vitro degradation study of the samples. Pure TPU shows the fastest degradation rate, while the slowest is that of pure PLA.

The biodegradation of polymeric biomaterials mainly depends on hydrolytic degradation of polymer chains. For example, degradation of polyurethane occurs when water molecules infiltrate the polymer network, triggering hydrolysis of the polyurethane chains, including the chemical dissolution of ester and amide bonds [76]. Similarly, the hydrolytic degradation of PLA started by the breaking of the ester link of the polymeric chain [77]. Hydrolytic degradation has been described as belonging to two types: surface erosion and bulk erosion. Surface erosion occurs exclusively at the polymer–water interface, while bulk erosion occurs uniformly throughout the polymer [76,77]. However, it is critical to emphasise that this degradation investigation was conducted on nonporous film specimens without plasma, biomolecules, or cells. The rate of degradation is important, as it is controlled by various other factors in vivo, which significantly enhances the degradation of polymers.

Erosion and bulk degradation are the two possible mechanisms through which polymers degrade. Crystallinity and chain orientation of polymers are particularly essential in degradation processes. The degradation process begins with water diffusion into the amorphous regions, followed by random hydrolytic scission of the ester bonds. After degrading the amorphous sections, the hydrolytic attack continues to the crystalline structures. Due to the loose packing of the molecules in the amorphous area of a polymer, they are more susceptible to attack by reactive species or solvents than the molecules in the crystalline region [78–80]. TPU's amorphous structure is the possible cause that results in a faster degradation rate than semicrystalline PLA, as seen by its lower glass transition temperature (T_g).

3.5. In Vitro Biocompatibility

The viability of cell culture is shown by the proportion of viable cells in a population. A decrease in the percentage of viable cells below 50% of total cell growth is described as the toxicity of the material towards the cells. BEAS-2B was cultured for three days on sterilised TPU, PLA, and TPU/PLA scaffolds to test their biocompatibility and toxicity of the material, and the result is shown in Figure 7.

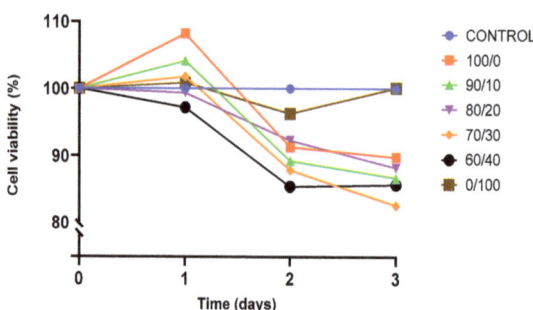

Figure 7. Presto Blue viability assay towards BEAS-2B cells.

Both pure TPU and PLA indicated biocompatibility towards BEAS-2B cells. After 24 h of incubation, cell proliferation was noted, exhibiting viability of cells was more than 100% in both pure materials. The result showed that the viability of BEAS-2B was greater than 80% in all compositions up until day 3, indicating that none of the compositions were toxic to the cells. Pure PLA outperformed pure TPU and other blends in terms of viability. TPU has been proven to be biocompatible and suitable for regenerative medicine, either alone or in conjunction with other polymers. Therefore, this result was expected [17,81,82]. This study is also consistent with the findings by Harynska and colleagues (2018) [83]. Similarly, the U.S. Food and Drug Administration (FDA) has authorised PLA as a biodegradable and biocompatible polymer for application in the human body due to its absorbability and nontoxicity [33,35,73,84].

4. Conclusions

Due to the shortage of organs and tissues for organ transplantation, synthetic materials are some of the treatment alternatives for trachea replacement, which necessitates developing an ideal material with desirable properties. Apart from biocompatibility and nontoxicity in the body system, mechanical properties are some of the important factors that must be considered due to the anatomy of the trachea in the body. TPU and PLA were selected for this study because of their inherent properties and wide use in biomedical applications. However, the properties of these materials limit their application. Polymer blending is an appealing and cost-effective method for developing new material with improved properties by combining physically existing polymers instead of synthesizing entirely new polymeric materials. This study investigated the mechanical properties, water absorption, biodegradability, and biocompatibility of melt blended TPU and PLA polymers as a material of choice in 3D printed tracheal tissue engineering. It was demonstrated that both materials were successfully compounded. Even though all blend compositions were not miscible, their morphology and mechanical properties showed improvement for the proposed use. Furthermore, the blended material is biocompatible and has an appropriate absorption and degradation rate, making it viable for use as a filament material in additive manufacturing for medical purposes. Therefore, the polymer blending of TPU and PLA can be a cost-effective strategy to improve the properties of TPU and PLA and produce filaments for 3D printing.

Author Contributions: Conceptualization, A.A.S. and B.H.Y.; methodology, A.A.S., B.H.Y., Z.A.A.H. and M.J.; validation, A.A.S., B.H.Y. and Z.A.A.H.; formal analysis, A.A.S., B.H.Y. and Z.A.A.H.; investigation, A.A.S.; resources, A.A.S., B.H.Y. and Z.A.A.H.; data curation, A.A.S. and B.H.Y.; writing—original draft preparation, A.A.S.; writing—review and editing, A.A.S., B.H.Y. and Z.A.A.H.; visualization, A.A.S., B.H.Y., Z.A.A.H. and M.J.; supervision, B.H.Y., Z.A.A.H. and M.J.; project administration, A.A.S. and B.H.Y.; funding acquisition, A.A.S. and B.H.Y. All authors have read and agreed to the published version of the manuscript.

Funding: This research was funded by the Ministry of Higher Education Malaysia for the Prototype Research Grant Scheme (PRGS) with Project Code PRGS/1/2021/SKK07/USM/02/1 and the Nippon Sheet Glass Foundation for Materials Science and Engineering (304/CIPPT/650932).

Institutional Review Board Statement: Not applicable.

Informed Consent Statement: Not applicable.

Data Availability Statement: Not applicable.

Acknowledgments: The authors would like to thank the Ministry of Higher Education Malaysia and the International Islamic University Malaysia for their support through the Academic Staff Training Scheme fellowship (ASTS) for Asmak Abdul Samat. The authors would like to thank to staff in Regenerative Medicine Laboratory, Advanced Medical and Dental Institute (IPPT), Universiti Sains Malaysia for their assistant and supports.

Conflicts of Interest: All authors declare no conflict of interest. The authors alone are responsible for the content and writing of the article.

References

1. Grillo, H.C. Tracheal replacement: A critical review. *Ann. Thorac. Surg.* **2002**, *73*, 1995–2004. [CrossRef]
2. Kucera, K.A.; Doss, A.E.; Dunn, S.S.; Clemson, L.A.; Zwischenberger, J.B. Tracheal replacements: Part 1. *ASAIO J.* **2007**, *53*, 497–505. [CrossRef] [PubMed]
3. US Health Resources and Services Administration. Organ Donation Statistics. Available online: https://www.organdonor.gov/statistics-stories/statistics.html#waiting-list (accessed on 2 December 2019).
4. Delaere, P.; Van Raemdonck, D. Tracheal replacement. *J. Thorac. Dis.* **2016**, *8*, S186–S196. [PubMed]
5. Chiang, T.; Pepper, V.; Best, C.; Onwuka, E.; Breuer, C.K. Clinical Translation of Tissue Engineered Trachea Grafts. *Ann. Otol. Rhinol. Laryngol.* **2016**, *125*, 873–885. [CrossRef]
6. Elliott, M.J.; De Coppi, P.; Speggiorin, S.; Roebuck, D.; Butler, C.R.; Samuel, E.; Crowley, C.; McLaren, C.; Fierens, A.; Vondrys, D.; et al. Stem-cell-based, tissue engineered tracheal replacement in a child: A 2-year follow-up study. *Lancet* **2012**, *380*, 994–1000. [CrossRef]
7. Hamilton, N.J.; Kanani, M.; Roebuck, D.J.; Hewitt, R.J.; Cetto, R.; Culme-Seymour, E.J.; Toll, E.; Bates, A.J.; Comerford, A.P.; McLaren, C.A.; et al. Tissue-Engineered Tracheal Replacement in a Child: A 4-Year Follow-Up Study. *Am. J. Transplant.* **2015**, *15*, 2750–2757. [CrossRef]
8. Townsend, J.M.; Ott, L.M.; Salash, J.R.; Fung, K.M.; Easley, J.T.; Seim, H.B.; Johnson, J.K.; Weatherly, R.A.; Detamore, M.S. Reinforced Electrospun Polycaprolactone Nanofibers for Tracheal Repair in an in Vivo Ovine Model. *Tissue Eng. Part A* **2018**, *24*, 1301–1308. [CrossRef] [PubMed]
9. Jing, X.; Mi, H.Y.; Salick, M.R.; Cordie, T.; Crone, W.C.; Peng, X.F.; Turng, L.S. Morphology, mechanical properties, and shape memory effects of poly(lactic acid)/thermoplastic polyurethane blend scaffolds prepared by thermally induced phase separation. *J. Cell. Plast.* **2014**, *50*, 361–379. [CrossRef]
10. Xia, D.; Jin, D.; Wang, Q.; Gao, M.; Zhang, J.; Zhang, H.; Bai, J.; Feng, B.; Chen, M.; Huang, Y.; et al. Tissue-engineered trachea from a 3D-printed scaffold enhances whole-segment tracheal repair in a goat model. *J. Tissue Eng. Regen. Med.* **2019**, *13*, 694–703. [CrossRef] [PubMed]
11. Gao, M.; Zhang, H.; Dong, W.; Bai, J.; Gao, B.; Xia, D.; Feng, B.; Chen, M.; He, X.; Yin, M.; et al. Tissue-engineered trachea from a 3D-printed scaffold enhances whole-segment tracheal repair. *Sci. Rep.* **2017**, *7*, 5246. [CrossRef]
12. O'Brien, F.J. Biomaterials & scaffolds for tissue engineering. *Mater. Today* **2011**, *14*, 88–95.
13. Sung, H.J.; Meredith, C.; Johnson, C.; Galis, Z.S. The effect of scaffold degradation rate on three-dimensional cell growth and angiogenesis. *Biomaterials* **2004**, *25*, 5735–5742. [CrossRef]
14. ISO/ASTM 52900. *Standard Terminology for Additive Manufacturing–General Principles–Terminology*; ASTM International: West Conshohocken, PA, USA, 2015; Volume 1, pp. 1–9. Available online: http://compass.astm.org/EDIT/html_annot.cgi?ISOASTM52900+15 (accessed on 5 September 2021).
15. Tofail, S.A.M.; Koumoulos, E.P.; Bandyopadhyay, A.; Bose, S.; O'Donoghue, L.; Charitidis, C. Additive manufacturing: Scientific and technological challenges, market uptake and opportunities. *Mater. Today* **2018**, *21*, 22–37. [CrossRef]
16. Bourell, D.; Kruth, J.P.; Leu, M.; Levy, G.; Rosen, D.; Beese, A.M.; Clare, A. Materials for additive manufacturing. *CIRP Ann. Manuf. Technol.* **2017**, *66*, 659–681. [CrossRef]
17. Jung, S.Y.; Lee, S.J.; Kim, H.Y.; Park, H.S.; Wang, Z.; Kim, H.J.; Yoo, J.J.; Chung, S.M.; Kim, H.S. 3D printed polyurethane prosthesis for partial tracheal reconstruction: A pilot animal study. *Biofabrication* **2016**, *8*, 045015. [CrossRef] [PubMed]
18. Gao, B.; Jing, H.; Gao, M.; Wang, S.; Fu, W.; Zhang, X.; He, X.; Zheng, J. Long-segmental tracheal reconstruction in rabbits with pedicled Tissue-engineered trachea based on a 3D-printed scaffold. *Acta Biomater.* **2019**, *97*, 177–186. [CrossRef] [PubMed]
19. Kim, I.G.; Park, S.A.; Lee, S.H.; Choi, J.S.; Cho, H.; Lee, S.J.; Kwon, Y.W.; Kwon, S.K. Transplantation of a 3D-printed tracheal graft combined with iPS cell-derived MSCs and chondrocytes. *Sci. Rep.* **2020**, *10*, 4326. [CrossRef] [PubMed]

20. She, Y.; Fan, Z.; Wang, L.; Li, Y.; Sun, W.; Tang, H.; Zhang, L.; Wu, L.; Zheng, H.; Chen, C. 3D Printed Biomimetic PCL Scaffold as Framework Interspersed with Collagen for Long Segment Tracheal Replacement. *Front. Cell Dev. Biol.* **2021**, *9*, 33. [CrossRef]
21. Xie, F.; Zhang, T.; Bryant, P.; Kurusingal, V.; Colwell, J.M.; Laycock, B. Degradation and stabilization of polyurethane elastomers. *Prog. Polym. Sci.* **2019**, *90*, 211–268. [CrossRef]
22. Williamson, M.R.; Black, R.; Kielty, C. PCL-PU composite vascular scaffold production for vascular tissue engineering: Attachment, proliferation and bioactivity of human vascular endothelial cells. *Biomaterials* **2006**, *27*, 3608–3616. [CrossRef]
23. Bergmeister, H.; Seyidova, N.; Schreiber, C.; Strobl, M.; Grasl, C.; Walter, I.; Messner, B.; Baudis, S.; Fröhlich, S.; Marchetti-Deschmann, M.; et al. Biodegradable, thermoplastic polyurethane grafts for small diameter vascular replacements. *Acta Biomater.* **2015**, *11*, 104–113. [CrossRef]
24. Yu, E.; Zhang, J.; Thomson, J.A.; Turng, L.S. Fabrication and characterization of electrospun thermoplastic polyurethane/fibroin small-diameter vascular grafts for vascular tissue engineering. *Int. Polym. Process.* **2016**, *31*, 638–646. [CrossRef] [PubMed]
25. Du, Y.J.; Brash, J.L.; McClung, G.; Berry, L.R.; Klement, P.; Chan, A.K.C.; Suwandi, J.S.; Toes, R.E.M.; Nikolic, T.; Roep, B.O. Protein adsorption on polyurethane catheters modified with a novel antithrombin-heparin covalent complex. *J. Biomed. Mater. Res. Part A* **2007**, *80*, 216–225. [CrossRef] [PubMed]
26. Kim, L.; Hermel-Davidock, T.; Weimer, M.W.; Burkolz, J.K. Catheter Tubing with Tailored Modulus Response. U.S. Patent 10596302B2, 17 April 2017.
27. Shie, M.Y.; Chang, W.C.; Wei, L.J.; Huang, Y.H.; Chen, C.H.; Shih, C.T.; Chen, Y.W.; Shen, Y.F. 3D printing of cytocompatible water-based light-cured polyurethane with hyaluronic acid for cartilage tissue engineering applications. *Materials* **2017**, *10*, 136. [CrossRef] [PubMed]
28. Ge, Z.; Li, C.; Heng, B.C.; Cao, G.; Yang, Z. Functional biomaterials for cartilage regeneration. *J. Biomed. Mater. Res. Part A* **2012**, *100A*, 2526–2536. [CrossRef] [PubMed]
29. Södergård, A.; Stolt, M. Properties of lactic acid based polymers and their correlation with composition. *Prog. Polym. Sci.* **2002**, *27*, 1123–1163. [CrossRef]
30. Casalini, T.; Rossi, F.; Castrovinci, A.; Perale, G. A Perspective on Polylactic Acid-Based Polymers Use for Nanoparticles Synthesis and Applications. *Front. Bioeng. Biotechnol.* **2019**, *7*, 259. [CrossRef]
31. Liu, S.; Yu, J.; Li, H.; Wang, K.; Wu, G.; Wang, B.; Liu, M.; Zhang, Y.; Wang, P.; Zhang, J.; et al. Controllable drug release behavior of polylactic acid (PLA) surgical suture coating with ciprofloxacin (CPFX)-polycaprolactone (PCL)/polyglycolide (PGA). *Polymers* **2020**, *12*, 288. [CrossRef]
32. Tappa, K.; Jammalamadaka, U.; Weisman, J.A.; Ballard, D.H.; Wolford, D.D.; Pascual-Garrido, C.; Wolford, L.M.; Woodard, P.K.; Mills, D.K. 3D printing custom bioactive and absorbable surgical screws, pins, and bone plates for localized drug delivery. *J. Funct. Biomater.* **2019**, *10*, 17. [CrossRef]
33. Sikhosana, S.T.; Gumede, T.P.; Malebo, N.J.; Ogundeji, A.O. Poly(Lactic acid) and its composites as functional materials for 3-d scaffolds in biomedical applications: A mini-review of recent trends. *Express Polym. Lett.* **2021**, *15*, 568–580. [CrossRef]
34. Saini, P.; Arora, M.; Kumar, M.N.V.R. Poly(lactic acid) blends in biomedical applications. *Adv. Drug Deliv. Rev.* **2016**, *107*, 47–59. [CrossRef] [PubMed]
35. Hamad, K.; Kaseem, M.; Ayyoob, M.; Joo, J.; Deri, F. Polylactic acid blends: The future of green, light and tough. *Prog. Polym. Sci.* **2018**, *85*, 83–127. [CrossRef]
36. Lin, T.A.; Lou, C.W.; Lin, J.H. The effects of thermoplastic polyurethane on the structure and mechanical properties of modified polypropylene blends. *Appl. Sci.* **2017**, *7*, 1254. [CrossRef]
37. Grimmer, J.F.; Gunnlaugsson, C.B.; Alsberg, E.; Murphy, H.S.; Kong, H.J.; Mooney, D.J.; Weatherly, R.A. Tracheal reconstruction using tissue-engineered cartilage. *Arch. Otolaryngol. Head Neck Surg.* **2004**, *130*, 1191–1196. [CrossRef] [PubMed]
38. Komura, M.; Komura, H.; Kanamori, Y.; Tanaka, Y.; Suzuki, K.; Sugiyama, M.; Nakahara, S.; Kawashima, H.; Hatanaka, A.; Hoshi, K.; et al. An animal model study for tissue-engineered trachea fabricated from a biodegradable scaffold using chondrocytes to augment repair of tracheal stenosis. *J. Pediatr. Surg.* **2008**, *43*, 2141–2146. [CrossRef]
39. Kang, N.; Liu, X.; Guan, Y.; Wang, J.; Gong, F.; Yang, X.; Yan, L.; Wang, Q.; Fu, X.; Cao, Y.; et al. Effects of co-culturing BMSCs and auricular chondrocytes on the elastic modulus and hypertrophy of tissue engineered cartilage. *Biomaterials* **2012**, *33*, 4535–4544. [CrossRef]
40. Romanova, O.A.; Tenchurin, T.H.; Demina, T.S.; Sytina, E.V.; Shepelev, A.D.; Rudyak, S.G.; Klein, O.I.; Krasheninnikov, S.V.; Safronova, E.I.; Kamyshinsky, R.A.; et al. Non-woven bilayered biodegradable chitosan-gelatin-polylactide scaffold for bioengineering of tracheal epithelium. *Cell Prolif.* **2019**, *52*, e12598. [CrossRef]
41. Omori, K.; Tada, Y.; Suzuki, T.; Nomoto, Y.; Matsuzuka, T.; Kobayashi, K.; Nakamura, T.; Kanemaru, S.; Yamashita, M.; Asato, R. Clinical application of in situ tissue engineering using a scaffolding technique for reconstruction of the larynx and trachea. *Ann. Otol. Rhinol. Laryngol.* **2008**, *117*, 673–678. [CrossRef] [PubMed]
42. Hamaji, M.; Kojima, F.; Koyasu, S.; Tsuruyama, T.; Komatsu, T.; Ikuno, T.; Date, H.; Nakamura, T. Development of a composite and vascularized tracheal scaffold in the omentum for in situ tissue engineering: A canine model. *Interact. Cardiovasc. Thorac. Surg.* **2014**, *19*, 357–362. [CrossRef]
43. Best, C.A.; Pepper, V.K.; Ohst, D.; Bodnyk, K.; Heuer, E.; Onwuka, E.A.; King, N.; Strouse, R.; Grischkan, J.; Breuer, C.K.; et al. Designing a tissue-engineered tracheal scaffold for preclinical evaluation. *Int. J. Pediatr. Otorhinolaryngol.* **2018**, *104*, 155–160. [CrossRef]

44. Haliloglu, T.; Onar, V.; Yildirim, G.; Sapci, T.; Savci, N.; Kahvecioglu, O.; Karavus, A. Tracheal reconstruction with porous high-density polyethylene tracheal prosthesis. *Ann. Otol. Rhinol. Laryngol.* **2000**, *109*, 981–987. [CrossRef]
45. Lin, C.H.; Su, J.M.; Hsu, S.H. Evaluation of type II collagen scaffolds reinforced by poly(ε-caprolactone) as tissue-engineered trachea. *Tissue Eng. Part C Methods* **2008**, *14*, 69–77. [CrossRef]
46. Rehmani, S.S.; Al-Ayoubi, A.M.; Ayub, A.; Barsky, M.; Lewis, E.; Flores, R.; Lebovics, R.; Bhora, F.Y. Three-Dimensional-Printed Bioengineered Tracheal Grafts: Preclinical Results and Potential for Human Use. *Ann. Thorac. Surg.* **2017**, *104*, 998–1004. [CrossRef] [PubMed]
47. Yin, H.; Wang, J.; Gu, Z.; Feng, W.; Gao, M.; Wu, Y.; Zheng, H.; He, X.; Mo, X. Evaluation of the potential of kartogenin encapsulated poly(L-lactic acid-co-caprolactone)/collagen nanofibers for tracheal cartilage regeneration. *J. Biomater. Appl.* **2017**, *32*, 331–341. [CrossRef] [PubMed]
48. Chang, C.S.; Yang, C.Y.; Hsiao, H.Y.; Chen, L.; Chu, I.M.; Cheng, M.H.; Tsao, C.K. Cultivation of auricular chondrocytes in poly(ethylene glycol)/poly(ε-caprolactone) hydrogel for tracheal cartilage tissue engineering in a rabbit model. *Eur. Cells Mater.* **2018**, *35*, 350–364. [CrossRef] [PubMed]
49. Safshekan, F.; Tafazzoli-Shadpour, M.; Abdouss, M.; Shadmehr, M.B. Mechanical characterization and constitutive modeling of human trachea: Age and gender dependency. *Mater. Basel* **2016**, *9*, 456. [CrossRef] [PubMed]
50. Dong, Z.; Li, Y.; Zou, Q. Degradation and biocompatibility of porous nano-hydroxyapatite/polyurethane composite scaffold for bone tissue engineering. *Appl. Surf. Sci.* **2009**, *255*, 6087–6091. [CrossRef]
51. Mi, H.Y.; Salick, M.R.; Jing, X.; Jacques, B.R.; Crone, W.C.; Peng, X.F.; Turng, L.S. Characterization of thermoplastic polyurethane/polylactic acid (TPU/PLA) tissue engineering scaffolds fabricated by microcellular injection molding. *Mater. Sci. Eng. C* **2013**, *33*, 4767–4776. [CrossRef] [PubMed]
52. Tang, Q.; Gao, K. Structure analysis of polyether-based thermoplastic polyurethane elastomers by FTIR, 1H NMR and 13C NMR. *Int. J. Polym. Anal. Charact.* **2017**, *22*, 569–574. [CrossRef]
53. Pan, R.; Yang, L.; Zheng, L.; Hao, L.; Li, Y. Microscopic morphology, thermodynamic and mechanical properties of thermoplastic polyurethane fabricated by selective laser sintering. *Mater. Res. Express* **2020**, *7*, 055301. [CrossRef]
54. Guo, Z.; Yang, C.; Zhou, Z.; Chen, S.; Li, F. Characterization of biodegradable poly(lactic acid) porous scaffolds prepared using selective enzymatic degradation for tissue engineering. *RSC Adv.* **2017**, *7*, 34063–34070. [CrossRef]
55. Yang, X.Z.; Wang, Y.C.; Tang, L.Y.; Xia, H.; Wang, J. Synthesis and characterization of amphophilic block copolymer of polyphosphoester and poly(L-lactic acid). *J. Polym. Sci. Part A Polym. Chem.* **2008**, *46*, 6425–6434. [CrossRef]
56. Oliaei, E.; Kaffashi, B.; Davoodi, S. Investigation of structure and mechanical properties of toughened poly(l-lactide)/thermoplastic poly(ester urethane) blends. *J. Appl. Polym. Sci.* **2016**, *133*. [CrossRef]
57. Jašo, V.; Cvetinov, M.; Rakič, S.S.S.; Petrovič, Z.S. Bio-plastics and elastomers from polylactic acid/thermoplastic polyurethane blends. *J. Appl. Polym. Sci.* **2014**, *131*. [CrossRef]
58. Focarete, M.L.; Scandola, M.; Dobrzynski, P.; Kowalczuk, M. Miscibility and mechanical properties of blends of (L)-lactide copolymers with atactic poly(3-hydroxybutyrate). *Macromolecules* **2002**, *35*, 8472–8477. [CrossRef]
59. Zhang, K.; Ran, X.; Wang, X.; Han, C.; Han, L.; Wen, X.; Zhuang, Y.; Dong, L. Improvement in toughness and crystallization of poly(L-lactic acid) by melt blending with poly(epichlorohydrin-co-ethylene oxide). *Polym. Eng. Sci.* **2011**, *51*, 2370–2380. [CrossRef]
60. Imre, B.; Bedo, D.; Domján, A.; Schön, P.; Vancso, G.J.; Pukánszky, B. Structure, properties and interfacial interactions in poly(lactic acid)/polyurethane blends prepared by reactive processing. *Eur. Polym. J.* **2013**, *49*, 3104–3113. [CrossRef]
61. Mahmud, M.S.; Buys, Y.F.; Anuar, H.; Sopyan, I. Miscibility, morphology and mechanical properties of compatibilized polylactic acid/thermoplastic polyurethane blends. In *Materials Today, Proceedings of the RAMM 2018, Penang Malaysia, 27–29 November 2018*; Elsevier Ltd.: Amsterdam, The Netherlands, 2019; Volume 17, pp. 778–786.
62. Wang, J.; Zhang, Y.; Sun, W.; Chu, S.; Chen, T.; Sun, A.; Guo, J.; Xu, G. Morphology Evolutions and Mechanical Properties of In Situ Fibrillar Polylactic Acid/Thermoplastic Polyurethane Blends Fabricated by Fused Deposition Modeling. *Macromol. Mater. Eng.* **2019**, *304*, 1900107. [CrossRef]
63. Roberts, C.R.; Rains, J.K.; Paré, P.D.; Walker, D.C.; Wiggs, B.; Bert, J.L. Ultrastructure and tensile properties of human tracheal cartilage. *J. Biomech.* **1998**, *31*, 81–86. [CrossRef]
64. Safshekan, F.; Tafazzoli-Shadpour, M.; Abdouss, M.; Shadmehr, M.B.; Ghorbani, F. Investigation of the mechanical properties of the human tracheal cartilage. *Tanaffos* **2017**, *16*, 107–114.
65. Wang, Y.; Liu, S.; Ding, K.; Zhang, Y.; Ding, X.; Mi, J. Quaternary Tannic Acid with Improved Leachability and Biocompatibility for Antibacterial Medical Thermoplastic Polyurethane Catheters. *J. Mater. Chem. B* **2021**, *9*, 4746–4762. Available online: https://pubs.rsc.org/en/content/articlehtml/2021/tb/d1tb00227a (accessed on 9 June 2021). [CrossRef] [PubMed]
66. Domínguez-Robles, J.; Mancinelli, C.; Mancuso, E.; García-Romero, I.; Gilmore, B.F.; Casettari, L.; Larrañeta, E.; Lamprou, D.A. 3D printing of drug-loaded thermoplastic polyurethane meshes: A potential material for soft tissue reinforcement in vaginal surgery. *Pharmaceutics* **2020**, *12*, 63. [CrossRef]
67. Lis-Bartos, A.; Smieszek, A.; Frańczyk, K.; Marycz, K. Fabrication, characterization, and cytotoxicity of thermoplastic polyurethane/poly(lactic acid) material using human adipose derived mesenchymal stromal stem cells (hASCs). *Polymers* **2018**, *10*, 1073. [CrossRef]

68. Thomas, S.; Shanks, R.; Chandrasekharakurup, S. *Design and Applications of Nanostructured Polymer Blends and Nanocomposite Systems*; Elsevier Inc.: Amsterdam, The Netherlands, 2015; ISBN 9780323394543.
69. Feng, F.; Ye, L. Morphologies and mechanical properties of polylactide/thermoplastic polyurethane elastomer blends. *J. Appl. Polym. Sci.* **2011**, *119*, 2778–2783. [CrossRef]
70. Rains, J.K.; Bert, J.L.; Roberts, C.R.; Pare, P.D. Mechanical properties of human tracheal cartilage. *J. Appl. Physiol.* **1992**, *72*, 219–225. [CrossRef] [PubMed]
71. Bonakdar, S.; Emami, S.H.; Shokrgozar, M.A.; Farhadi, A.; Ahmadi, S.A.H.; Amanzadeh, A. Preparation and characterization of polyvinyl alcohol hydrogels crosslinked by biodegradable polyurethane for tissue engineering of cartilage. *Mater. Sci. Eng. C* **2010**, *30*, 636–643. [CrossRef]
72. Gogolewski, S.; Gorna, K.; Zaczynska, E.; Czarny, A. Structure-property relations and cytotoxicity of isosorbide-based biodegradable polyurethane scaffolds for tissue repair and regeneration. *J. Biomed. Mater. Res. Part A* **2008**, *85*, 456–465. [CrossRef]
73. Revati, R.; Majid, M.S.A.; Ridzuan, M.J.M.; Basaruddin, K.S.; Rahman, Y.M.N.; Cheng, E.M.; Gibson, A.G. In vitro degradation of a 3D porous Pennisetum purpureum/PLA biocomposite scaffold. *J. Mech. Behav. Biomed. Mater.* **2017**, *74*, 383–391. [CrossRef]
74. Chen, Q.; Bruyneel, A.; Clarke, K.; Carr, C.; Czernuszka, J. Collagen-Based Scaffolds for Potential Application of Heart Valve Tissue Engineering. *J. Tissue Sci. Eng.* **2012**, *11*, 1–5. [CrossRef]
75. Tian, H.; Tang, Z.; Zhuang, X.; Chen, X.; Jing, X. Biodegradable synthetic polymers: Preparation, functionalization and biomedical application. *Prog. Polym. Sci.* **2012**, *37*, 237–280. [CrossRef]
76. Brzeska, J.; Heimowska, A.; Sikorska, W.; Jasińska-Walc, L.; Kowalczuk, M.; Rutkowska, M. Chemical and Enzymatic Hydrolysis of Polyurethane/Polylactide Blends. *Int. J. Polym. Sci.* **2015**, *2015*, 795985. [CrossRef]
77. Araque-Monrós, M.C.; Vidaurre, A.; Gil-Santos, L.; Gironés Bernabé, S.; Monleón-Pradas, M.; Más-Estellés, J. Study of the degradation of a new PLA braided biomaterial in buffer phosphate saline, basic and acid media, intended for the regeneration of tendons and ligaments. *Polym. Degrad. Stab.* **2013**, *98*, 1563–1570. [CrossRef]
78. Yoo, E.S.; Im, S.S. Effect of crystalline and amorphous structures on biodegradability of poly(tetramethylene succinate). *J. Environ. Polym. Degrad.* **1999**, *7*, 19–26. [CrossRef]
79. Mondal, S.; Martin, D. Hydrolytic degradation of segmented polyurethane copolymers for biomedical applications. *Polym. Degrad. Stab.* **2012**, *97*, 1553–1561. [CrossRef]
80. Elsawy, M.A.; Kim, K.H.; Park, J.W.; Deep, A. Hydrolytic degradation of polylactic acid (PLA) and its composites. *Renew. Sustain. Energy Rev.* **2017**, *79*, 1346–1352. [CrossRef]
81. Kucinska-Lipka, J.; Gubanska, I.; Strankowski, M.; Cieśliński, H.; Filipowicz, N.; Janik, H. Synthesis and characterization of cycloaliphatic hydrophilic polyurethanes, modified with L-ascorbic acid, as materials for soft tissue regeneration. *Mater. Sci. Eng. C* **2017**, *75*, 671–681. [CrossRef] [PubMed]
82. Borkenhagen, M.; Stoll, R.C.; Neuenschwander, P.; Suter, U.W.; Aebischer, P. In vivo performance of a new biodegradable polyester urethane system used as a nerve guidance channel. *Biomaterials* **1998**, *19*, 2155–2165. [CrossRef]
83. Haryńska, A.; Gubanska, I.; Kucinska-Lipka, J.; Janik, H. Fabrication and characterization of flexible medical-grade TPU filament for Fused Deposition Modeling 3DP technology. *Polymers* **2018**, *10*, 1304. [CrossRef] [PubMed]
84. Santoro, M.; Shah, S.R.; Walker, J.L.; Mikos, A.G. Poly(lactic acid) nanofibrous scaffolds for tissue engineering. *Adv. Drug Deliv. Rev.* **2016**, *107*, 206–212. [CrossRef]

Article

Biodegradable PGA/PBAT Blends for 3D Printing: Material Performance and Periodic Minimal Surface Structures

Zihui Zhang [1,†], Fengtai He [2,†], Bo Wang [3], Yiping Zhao [2], Zhiyong Wei [4], Hao Zhang [5,*] and Lin Sang [1,*]

1. School of Automotive Engineering, Dalian University of Technology, Dalian 116024, China; zhangzihui@mail.dlut.edu.cn
2. Department of Radiology, Second Affiliated Hospital of Dalian Medical University, Dalian 116027, China; hft19940214@163.com (F.H.); emmazhaochina@163.com (Y.Z.)
3. School of Materials Science and Engineering, Dalian University of Technology, Dalian 116024, China; BobWang@mail.dlut.edu.cn
4. Department of Polymer Science and Engineering, School of Chemical Engineering, Dalian University of Technology, Dalian 116024, China; zywei@dlut.edu.cn
5. Department of Orthopedics, Affiliated Dalian Municipal Central Hospital, Dalian Medical University, Dalian 116027, China
* Correspondence: zhanghao20201208@163.com (H.Z.); sanglin@dlut.edu.cn (L.S.)
† These authors contributed equally to this work.

Abstract: Biodegradable polymers have been rapidly developed for alleviating excessive consumption of non-degradable plastics. Additive manufacturing is also a green energy-efficiency and environment-protection technique to fabricate complicated structures. Herein, biodegradable polyesters, polyglycolic acid (PGA) and poly (butyleneadipate-co-terephthalate) (PBAT) were blended and developed into feedstock for 3D printing. Under a set of formulations, PGA/PBAT blends exhibited a tailored stiffness-toughness mechanical performance. Then, PGA/PBAT (85/15 in weight ratio) with good thermal stability and mechanical property were extruded into filaments with a uniform wire diameter. Mechanical testing clearly indicated that FDM 3D-printed exhibited comparable tensile, flexural and impact properties with injection-molded samples of PGA/PBAT (85/15). Furthermore, uniform and graded Diamond-Triply Periodic Minimal Surfaces (D-TPMS) structures were designed and successfully manufactured via the fused deposition modeling (FDM) technique. Computer tomography (CT) was employed to confirm the internal three-dimensional structures. The compressive test results showed that PGA/PBAT (85/15) D-surface structures bear better load-carrying capacity than that of neat PGA, giving an advantage of energy absorption. Additionally, typical industrial parts were manufactured with excellent dimension-stability, no-wrapping and fine quality. Collectively, biodegradable PGA/PBAT material with good printability has great potentials in application requiring stiffer structures.

Keywords: biodegradable polyesters; polyglycolic acid (PGA); fused deposition modeling (FDM); triply periodic minimal surfaces (TPMS); mechanical property

1. Introduction

In recent years, fused deposition modeling (FDM) has been developed rapidly among 3D printing techniques because of its low cost in maintenance, and diversity in thermoplastic feedstock [1–3]. It enables the production of custom parts with complex structures in many application fields including the medical, food, automotive, aerospace and construction industries [4–6]. Among the commercial thermoplastic materials used in FDM 3D printing, poly (lactic acid) (PLA) is rather popular in 3D printing because of its biodegradability, biocompatibility, favorable mechanical properties and facile printability. Nevertheless, PLA-based materials have inherent limitations of brittleness and low toughness [7,8]. With the request of environmental protection and the growing demands in biodegradable polymers, considerable interest has been attracted to develop various biodegradable polymers as feedstock for 3D printing [9–12].

Polyglycolic acid (PGA), a biodegradable polymer, can be degraded to carbon dioxide and water with a relatively fast degradation rate [13–15]. Its degradation products can be absorbed by the human body, which is approved by US Food and Drug Administration (UFDA). Besides, PGA also has superior mechanical strength to other biodegradable polymers [16,17]. It is found that the mechanical strength and modulus are similar to that of human bones, which make it an ideal candidate for hard tissue implanted materials. Wu et al. [18] fabricated polyetheretherketone (PEEK)/polyglycolide acid (PGA) scaffolds using 3D printing technology showed the ability to efficiently sustain drug release as an implant for treating bacterial infection. Taegyun et al. [19] 3D-printed PGA/hydroxyapatite composite scaffolds and demonstrated it can promote patient-specific bone regeneration. The existed literature mainly reported on the incorporation of PGA with other plastics or bioactive fillers, however, few studies focused on the PGA-based filaments for FDM 3D printing and their printability and precision of complex structures.

Currently, biodegradable elastomeric polyesters, such as poly (butyleneadipate-co-terephthalate) (PBAT) [20], poly (butylene succinate) (PBS) [21] and polycaprolactone (PCL) [22] have developed vigorously. It is hoped that blending PGA with these ductile polymers will develop 3D-printing feedstock with good biocompatibility, tailored biodegradability and mechanical properties. Among them, PBAT possesses an excellent ductile property with high elongation at break. Therefore, PBAT is considered a good candidate to improve the flexibility of PGA. It is expected that PGA/PBAT blends could be novel 3D-printing feedstock due to the high mechanical strength of PGA and the elevated toughness of PBAT.

Although there was little evidence of PGA processed by FDM printing in the past, extruding, sintering and injection molding has been attempted [23–25]. Additionally, PBAT-based materials (i.e., PBAT/PLA) have also been fabricated into 3D-printing feedstock [26,27]. Therefore, binary PGA/PBAT blends for FDM 3D printing are proposed and subjected to this research. An important aspect of this system is the phase compatibility of the blend, which would directly influence the mechanical performance. Thus, an epoxy-functionalized chain extender was used to react with the carboxyl and carboxyl functional groups in both polyesters [28,29].

The focus of this article is to explore the feasibility of novel PGA/PBAT blends as 3D printing feedstock and manufacturing complex lattice structures with good quality. Firstly, different compositions of PGA/PBAT blends were compounded, the thermal behavior, thermal stability and mechanical performance were subsequently evaluated and the optimization of PGA/PBAT blends for FDM printing was obtained. Then, the emphasis on fabricating PGA/PBAT filaments was carried out, and the mechanical properties of FDM-printed and injection-molded samples were comprehensively evaluated. Finally, periodic minimal surface structures with constant-thickness and graded-thickness were designed and manufactured, and the corresponding compressive performance was compared with the pure PGA group. In summary, this work is to explore the potentials of novel biodegradable PGA/PBAT filament in complex shape, high strength, and lightweight engineering applications in a sustainable and energy conservation way.

2. Materials and Methods
2.1. Materials

Commercial PBAT having a density of 1.21 g/cm^3, melting temperature of 125 °C and a melt flow rate (MFR) rate of 44 g/10 min (230 °C and 2.16 kg) was obtained from Kanghui New Materials Hi-Tech Co., Ltd. (Yingkou, China). PGA having a density of 1.64 g/cm^3, melting temperature of 220 °C and a melt flow rate (MFR) rate of 40 g/10 min (230 °C and 2.16 kg) was kindly provided by Shanghai Pujing Chemical Industry Co., Ltd. (Shanghai, China). Both PGA and PBAT were fully biodegradable polyesters, and the chemical structures were given in Figure 1. A multi-functional epoxy chain extender styrene-glycidyl methacrylate (Joncryl ADR 4370) was purchased from BASF Chemical Company (Ludwigshafen, Germany).

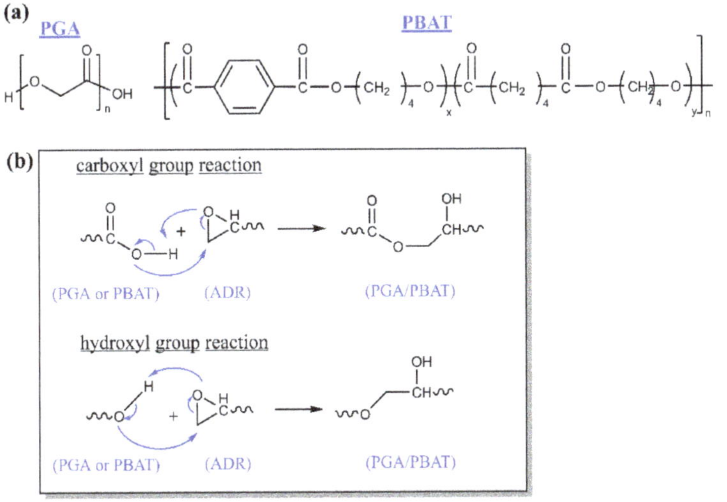

Figure 1. (a) Chemical structure of PGA and PBAT, (b) Reaction mechanism between PGA and PBAT with ADR.

2.2. Sample Preparation

Before the extrusion processing, the PBAT pellets were vacuum-dried at 80 °C for 8 h to remove the moisture. PGA pellets were kept in vacuum-sealed bags with a desiccant at 4 °C and dried at 60 °C for 2 h. In order to avoid undesirable hydrolysis during extruding, the predrying process was conducted to remove moisture. Joncryl ADR was used as received. The mass formulations of PGA/PBAT composites contained 100/0, 95/5, 85/15 and 75/15, and the content of Joncryl ADR was 1.5 wt% of the whole biodegradable polyesters. The PGA and PBAT pellets were compounded by a twin-screw extruder (SHJ20-X40, L/D = 40, D = 40 mm, Nanjing Giant Machinery Co., Ltd., Nanjing, China). The processing temperatures of extruding zones were set from 210–230 °C with a rotation speed of 40 rpm.

Test specimens for the tensile test were molded using injection molding equipment (Wuhan Ruiming Machinery Co., Ltd. Wuhan, China) for each blend composition (PGA/PBAT: 100/0, 95/5, 85/15, 75/25). The heating zone of the injection molding system was 230 °C and the molding zone was 40 °C. The standard dumbbell-shaped specimens (ISO 527, type2) were prepared and stored in a sealed dryer before characterization.

2.3. Filament Feedstock Fabrication and 3D Printing

Under the optimized formulations of PGA/PBAT composites, PGA/PBAT (85/15) was adopted to fabricate into filaments (1.75 ± 0.5 mm diameter) using a desktop single-screw filament extruder (Wellzoom C, Shenzhen Mistar Technology Co., Ltd., Shenzhen, China). The extruder barrel heating zone to die temperature was set at 225 and 230 °C, respectively. The extruded filaments were pulled through and collected by a winding unit. The filaments were dried in an air-circulating oven at 60 °C for 4 h and stored in sealed vacuumed bags with desiccant (4 °C) prior to 3D printing or other characterization.

3D-printed PGA/PBAT samples were manufactured via an FDM 3D printer (FUNMAT HT, INTAMSYS Co. Ltd., Xi'an, China). The FDM printing condition was set as: a nozzle diameter of 0.40 mm, nozzle temperature at 230 °C, as building platform temperature at 45 °C, ambient temperature at 45 °C, infill density of 100%, printing speed maintained at 20 mm/s, raster angle of 45° and each layer thickness of 0.10 mm.

2.4. Design of the Diamond-Triply Periodic Minimal Surfaces (D-TPMS) Structures

Triply periodic minimal surfaces (TPMS) can be mathematically approximated using implicit methods [30,31]. Among types of TPMS structures, Diamond (D) surface was selected in the current work. The D-surface is described as follows:

$$\phi D(x,y,z) = \sin(\omega)\sin(\omega y)\sin(\omega z) + \cos(\omega x)\sin(\omega y)\sin(\omega z) + \\ \sin(\omega x)\cos(\omega y)\sin(\omega z) + \sin(\omega x)\sin(\omega y)\cos(\omega z) = C \quad (1)$$

where x, y, z represent spatial coordinates, $w = 2\pi/l$ and l is the length of a unit cell. The 3D D-surface is generated as the solution of the level-set function $\phi = C$. The solid model of Diamond surfaces was created by extracting the zero-level set surface (when C = 0) from Equation (1). Matlab scripting was used to generate the sheet surfaces. The total cylinder sample size has a diameter of 20 mm and a height of 20 mm. Constant-thickness of structures with nominal average wall thickness was 0.4 mm and 2 mm, respectively. The wall thickness of graded structures ranged from 0.4 mm to 2 mm radically. The resultant 3D stereolithography (STL) models were then transferred to CURA software (Ultimaker Co. Ltd., Amsterdam, Holland) for slicing in preparation for 3D printing.

2.5. Characterization

Differential scanning calorimeter (DSC). The non-isothermal crystallization and melting behavior of PGA and PGA/PBAT were studied under a nitrogen atmosphere (10 °C/min). The weight of test samples was 5–8 mg and sealed in an aluminum pan. The running program was divided into three stages: heating room temperature to 250 °C, annealing at 250 °C for 5 min; cooling to 20 °C, maintaining for 3 min; reheating to 250 °C.

Thermogravimetric analysis (TGA). Thermal decomposition stability of PGA and PGA/PBAT composites was carried out through TGA (Q600, TA instruments, NewCastle, America). The heating program was from 30 °C to 600 °C under a nitrogen atmosphere at a heating rate of 10 °C/min.

Scanning electron microscopy. The compatibility of PGA and PBAT in injection-molded and 3D-printed samples was observed by a tungsten filament scanning electron microscope (SEM, QUANTA 450, FEI, Hillsboro, America). The fractured surface was spray-covered with a thin gold layer.

2.6. Mechanical Test

The mechanical properties of injection-molded and FDM-printed samples were tested by using a mechanical testing machine (GT-7001-HC6, GOTECH TESTING MACHINE INC, Taiwan, China). The tensile strength and modulus were measured at a constant crosshead speed of 2 mm/min at an ambient temperature according to the standard of ISO 527. For flexural tests according to ISO 178 standard, three-point bending with a crosshead speed of 2 mm/min was performed at room temperature.

The stiffness of the FDM-printed D-surface TPMS structures was evaluated from compression tests, according to the ASTM D-695 standard. The unconstrained cellular structure samples were 20 mm in diameter and the height was 20 mm, which were compressed between two rigid flat steel plates with a constant strain rate of 2 mm/min. The compressive force and displacement data from the universal machine were recorded. When the strain of cellular structures reached 20%, the tests were terminated.

The total energy absorption (EA) was calculated from the area under the force-displacement curve as follows, Equation (2) [32,33]:

$$EA = \int_a^b \sigma \cdot d\varepsilon \quad (2)$$

where σ assigns to the compressive stress and the ε is the nominal strain. The calculation of $\sigma = F/A$ and $\varepsilon = \delta/H$, respectively. F and δ corresponded to the compressive force and displacement, which are recorded during the compression test. A is the original cross-section area and H is the height of the D-surface TPMS structure along the compressive direction.

2.7. Computed Tomography

The deformation of the D-surface TPMS structure samples after compression was scanned by Computed tomography (CT equipment, SIEMENS SOMATOM DRIVE, langen, Germany). CT scan conditions as follows: the voltage was 70 kV, the current was 61 mA, slice thickness was 0.5 mm, slice spacing was 0.3 mm, FOV (field of view) was 50 mm, the matrix was 512 × 512, and DLP (dose length product) was approximately 45.58 mGy. The obtained scan data were subsequently reconstructed using SIMENS software analysis system (Syngo CT VA62A, Erlangen, Germany).

3. Results and Discussion
3.1. Preparation and Characterization of PGA/PBAT Samples

The epoxy group of ADR is expectable to react with both hydroxyl and carboxyl of the polyester [34,35]. The introduction of ADR efficiently facilitated the reaction between PGA and PBAT, which resulted in a polymer network (as illustrated in Figure 1) and decreased both the number of hydroxyl and carboxyl end-groups in PGA and PBAT. The cross-linking reaction would increase the melt strength and protect the sensitive groups from hydrolysis degradation during the extruding process.

Prior to filament fabrication and FDM printing, thermal behavior and stability of filaments are essential due to the necessary information on the printed component of the printing window. The DSC thermograms of PGA, PGA/PBAT (95/5), PGA/PBAT (85/15) and PGA/PBAT (75/25) were shown in Figure 2a, b. After eliminating thermal history, the data in the first cooling and second heating scan were collected and plotted into curves. The values of crystallization temperature (T_c), the crystallization enthalpy (ΔH_c) in the first cooling curves; the melting temperature (T_m) and melt enthalpy (ΔH_m) in the second heating curves were summarized in Table 1. It can be seen that T_c of the PBAT (~74 °C) was not detected from the cooling curves [36], while T_c of PGA (~190 °C) was obvious [37]. The crystallization peaks were related to the addition of PBAT, the T_c migrates to higher temperatures from 185.1 to 194.5 °C, which can be assigned to the heterogeneous nucleation effect of the branch chain of PBAT for the crystallization of PGA. However, the ΔH_c was continuously decreased when the content of PBAT reached 25 wt%. This was because that small content of PBAT acted as a dispersed phase while PGA was the continuous phase. The increased content of entangled chains of PBAT might influence the organized PGA polymer crystallization. In the second heating curve, there is only one melting peak at 220.5 °C for neat PGA (100/0), whereas peaks are split into two in PGA/PBAT blends. With the increased content of PBAT, two melting peaks became evident. This was probably due to the formation of two crystalline structures of PGA co-existing in the binary blends.

Furthermore, weight loss and differential thermogravimetric curve (DTG) curves of neat PGA and PGA/PBAT blends were plotted in Figure 2c,d. For neat PGA, the loss weight curve showed that the weight started to decrease at 339.2 °C, and weight loss was quite obvious at 389.7 °C. After incorporation of 5 wt% and 15 wt% PBAT, the initial decomposition temperature slightly increased to 358.2 °C and 346.7 °C, which was probably due to enhanced stability by the cross-linking reaction using ADR. The fastest decomposition temperatures were also improved for PGA/PBAT (95/5) and PGA/PBAT (85/15) (as shown in Table 1). However, when the content of PBAT achieved 25 wt%, the characteristic decomposition peak of PBAT could be detected, and thus the decomposition temperature shifted to a lower temperature at 326.0 °C. Two remarkable decomposition stages were observed at 351.2 °C and 412.5 °C in PGA/PBAT (75/25). The first stage corresponded to the decomposition of PBAT, and the latter peak was assigned to the decomposition of PGA. These results illustrated that the thermal degradation of PGA/PBAT (95/5) and PGA/PBAT (85/15) could be effectively postponed after the extruding process in the presence of ADR, which was suitable as filament feedstock.

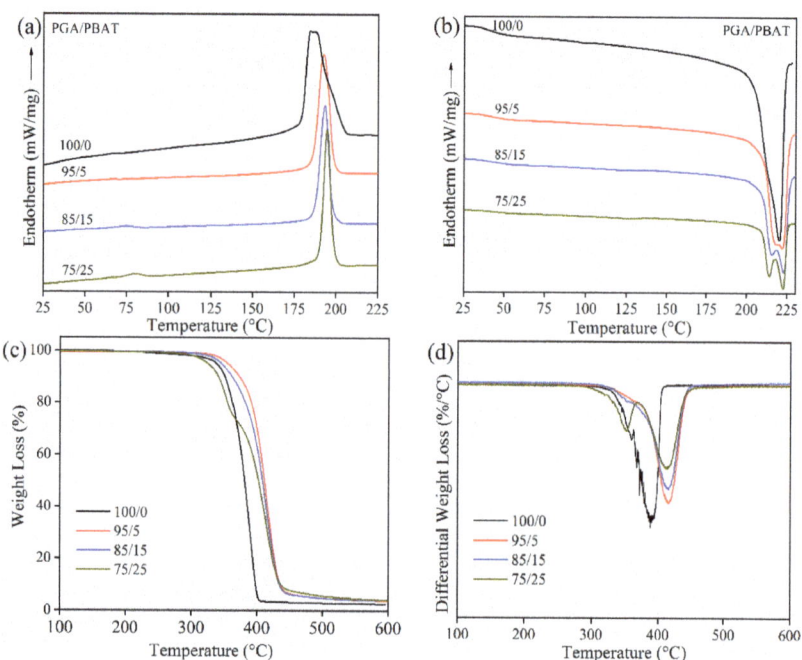

Figure 2. (**a**) Cooling; (**b**) Second heating DSC curves; (**c**) Weight loss; (**d**) Differential thermogravimetric curves of PGA and PGA/PBAT composite pellets.

Table 1. Crystalline and melting and decomposition parameters of PGA, PGA/PBAT (95/5), PGA/PBAT (85/15) and PGA/PBAT (75/25).

Specimen	T_c (°C)	ΔH_c (J/g)	T_m (°C)	ΔH_m (J/g)	$T_{d,5\%}$ (°C)	$T_{d,max}$ (°C)
PGA	185.1	76.3	220.5	84.5	339.2	389.7
PGA/PBAT(95/5)	192.0	64.7	217.7/222.2	68.3	358.2	415.7
PGA/PBAT(85/15)	193.3	58.8	215.5/222.3	61.8	346.7	414.4
PGA/PBAT(75/25)	194.5	51.2	214.2/222.4	51.5	326.0	412.5

$T_{d,5\%}$: the initial decomposition temperature, 5% of loss weight; $T_{max,5\%}$: temperature to the maximum decomposition rate.

The tensile properties of injection-molded (IM) PGA/PBAT with different formulations were presented in Figure 3. The results showed that the tensile strength and Young's modulus gradually decreased with the increase in PBAT content. The tensile strength of neat PGA was 114 ± 1.24 MPa and Young's modulus was 5.15 ± 0.15 GPa, which was even stiffer to poly(ether-ether-ketone) (PEEK) materials with a tensile strength of 100 MPa and Young's modulus of 3.7 GPa [38]. However, excess strength might bring difficulties (i.e., brittleness) in plastic processing and application fields. Since PBAT was an excellent elastomeric "soft" polyester, although its tensile strength was 18 MPa, the modulus was 800 MPa and the elongation at break could reach to ~800%. Therefore, in the current work, PBAT first attempted to modify PGA to broaden the applications. From Figure 3a, the tensile strength gradually decreased from 114 MPa (PGA) to 79 MPa (95/5), 60 MPa (85/15) and 45 MPa (75/25), respectively. On the contrary, the elongation at break was 2.3%, 4.5%, 15.6% and 20.2% for PGA (100/0), PGA/PBAT (95/5), PGA/PBAT (85/15) and PGA/PBAT (75/25), respectively (Figure 3b). Accordingly, although the stiffness of PGA/PBAT was weakened to some extent, elongation at break was significantly increased after incorporation with elastomeric PBAT.

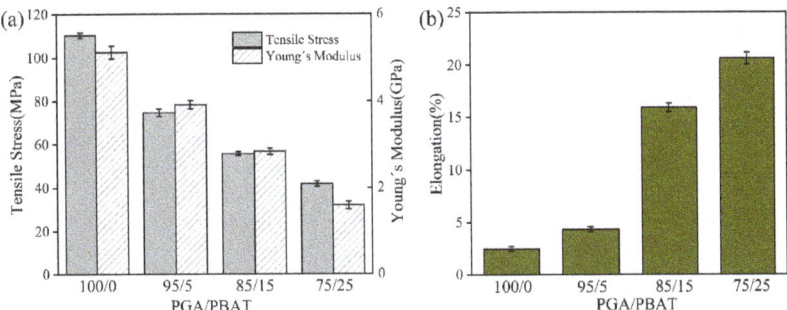

Figure 3. (a) Tensile strength, Young's modulus; (b) elongation at break of PGA/PBAT (100/0, 95/5, 85/15 and 75/25) blends.

To date, a great variety of synthetic polymers have been developed and used as feedstock materials, such as ABS, PA6, POM and PLA. Among them, PLA was the only commercial biodegradable filament, while other kinds of biodegradable polymer feedstocks were limited. In our work, three formulations of PGA/PBAT blends were fabricated. Among these formulations, the tensile strength and modulus of PGA/PBAT (85/15) were comparable with PLA (tensile strength: 65 MPa, modulus: 2.1 GPa) [39]. Moreover, its ductility was improved for the incorporation of PBAT, which was superior to neat PLA. In previous researches, some elastomeric polymers such PBS, PBAT and other polyesters were also applied to blend with PLA to overcome the inherent brittleness [40]. However, the mechanical properties of binary PLA-based materials were inferior to PGA/PBAT (85/15) prepared in the current work. Accordingly, PGA/PBAT (85/15) was supposed to be a good candidate material as a 3D-printing feedstock with balanced strength and toughness.

3.2. PGA/PBAT Filament Feedstock via 3D Printing

Based on the above analysis, neat PGA and PGA/PBAT (85/15) blends were fabricated in 3D printing filaments through a single screw extruder. As shown in Figure 4, uniform and standard filaments with a 1.75 mm diameter were obtained. According to T_m in Table 1, the temperature nozzle at 230 °C could successfully print the PGA/PBAT into the tensile specimens without any defects. The good printability of PGA/PBAT showed great potentials in the additive manufacturing of complicated parts.

Mechanical properties of the 3D-printed PGA/PBAT (85/15) specimens were comprehensively evaluated to compare with injection-molded (IM) control groups (Figures 5–7). The tensile, flexural and notched impact tests of the PGA/PBAT (85/15) specimens processed by injection molding and 3D printing were conducted. In Figure 5a, the tensile stress-strain curves of IM and 3D-printed samples showed a similar trend with an obvious necking stage. The elongation at breakage was all above 10%, and the value was higher in IM specimens. Ductile behavior with necking characteristics was observed, suggesting an improved toughness due to PBAT incorporation. In Figure 5b, the tensile strength and Young's modulus of 3D-printed specimens reached 56.5 MPa and 2.48 GPa respectively, which were equivalent to 94% and 85% of the injection-molded groups, respectively. A slightly lower value in the 3D-printed groups was determined by the layer-by-layer deposition mode. The delamination failure mode between printing layers may obstruct the force transfer when suffering from the external loading force.

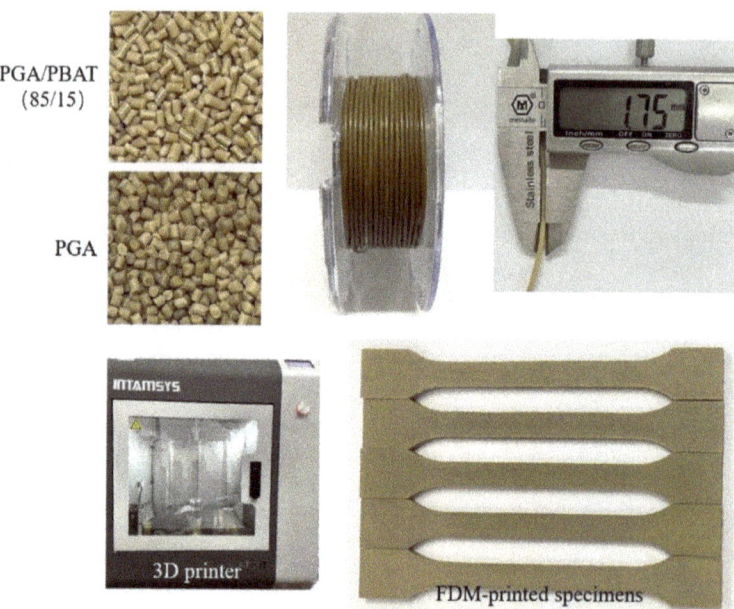

Figure 4. Original PGA and PGA/PBAT (85/15) composite pellets, extruded filaments of PGA/PBAT (85/15) and FDM-printed tensile specimens.

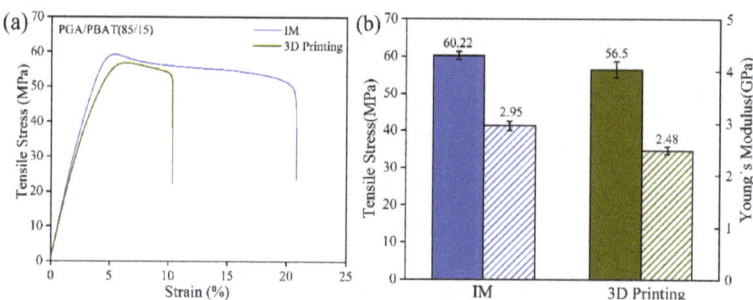

Figure 5. (**a**) Tensile stress-strain curves; (**b**) histograms of tensile strength and Young's modulus of PGA/PBAT (85/15) specimens fabricated by IM and 3D printing.

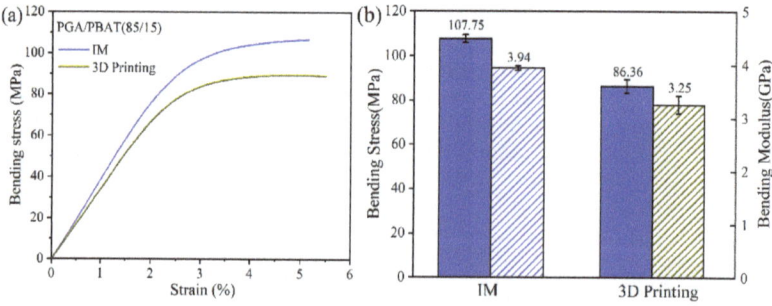

Figure 6. (**a**) Flexural stress-strain curves; (**b**) histograms of bending strength and modulus of PGA/PBAT (85/15) specimens fabricated by IM and 3D printing.

Figure 7. (**a**) Notch impact values; (**b**) fracture surfaces of PGA/PBAT (85/15) specimens fabricated by injection molding and 3D printing.

Similar results were seen in flexural and impact results (as shown in Figures 6 and 7) Compared with tensile performance, the difference between 3D-printed and IM groups flexural properties was slightly evident. The flexural strength and modulus of 3D-printed specimens were 80% and 82% of those of IM counterparts. This was mainly because the printing direction was vertical from the loading, which brought a challenge for the interface between the printing layers deposition technique. This layer-by-layer limitation was intensified using crystalline or semi-crystalline polymers, which might shrink or delaminate during the printing process. In Table 1, the enthalpy of crystallization was decreased with the increasing addition of PBAT, suggesting an inhibition effect of PBAT on PGA crystallization. Therefore, compounding blends with entangled or network molecular chains contributed to an enhanced interlayer quality for the FDM technique.

In order to further analyze the failure mechanism, the impact fracture surfaces of injection-molded and 3D-printed specimens were observed (Figure 7). It can be seen that ductile fracture occurred in both samples, and the fractured surface was partly concavo-convex. A few pores were observed in the 3D-printed sample surface (marked with a red circle), which was caused by layer-by-layer deposition during the printing process. Previous studies have demonstrated that the mechanical properties of 3D-printed samples could obtain ~80% of the injection-molded counterparts, which was restricted by the left voids or interface between the adjacent printing layers. Therefore, it is acceptable for slightly lower values of mechanical properties for PGA/PBAT (85/15) blend feedstock in comparison with IM materials. Nevertheless, the 3D-printed PGA/PBAT (85/15) filament possesses great potentials in fabricating complex structures.

3.3. Applications for PGA/PBAT Structure Manufacturing

It was demonstrated that TPMS structures possessed excellent energy absorption capacity, and the graded-thickness samples could avoid large stress fluctuations and show high cumulative energy absorption values than the constant-thickness samples [41,42]. Among the known TPMS structures (i.e., Diamond (D), Gyroid (G), I-WP, etc.), the stiffness, yield strength, ultimate strength and energy absorption capacity were investigated [43]. Results indicated that the TPMS-D structure exhibited excellent compressive property. Therefore, uniform and graded TPMS structures of D surfaces were adopted in this study. Images of geometric D-TPMS models, 3D-printed samples and the CT-reconstruction were assembled in Figure 8. Uniform pore architecture with two sheet thicknesses and radially graded structures were designed in the current work. The 3D-printed samples were highly coincident with geometric models, suggesting FDM printing was able to fabricate complex D-TPMS structures. Furthermore, 3D-reconstructed CT images obtained the three-dimension structure, confirming the desired structure was achieved by the FDM printing using PGA/PBAT (85/15). All samples possessed a well-controlled 3D porous structure with high interconnectivity. In 2D-reconstructed CT images, the cross-section images in the x-z and x-y planes clearly showed the internal pore architecture. In the constant-thickness D-TPMS structure, the sheet thickness was quite uniform, whereas the thickness from thin to thick was radially distributed in the cylinder.

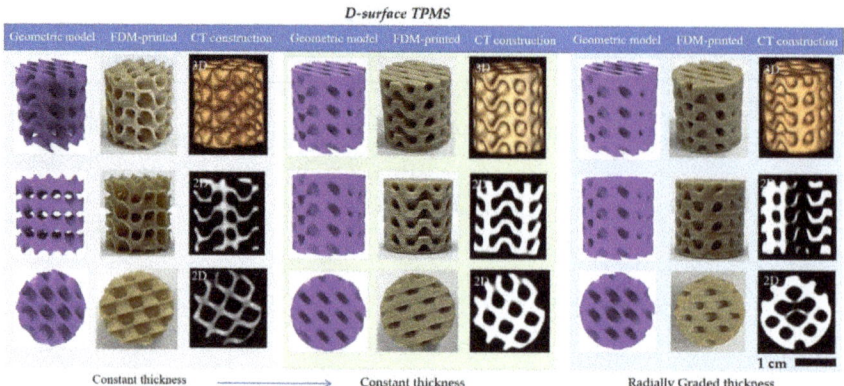

Figure 8. Geometric models, 3D-printed and CT reconstruction images of D-surface TPMS structures with the uniform and radially graded pore architectures.

The stress-strain curves of the printed D-surfaces of neat PGA and PGA/PBAT (85/15) were presented in Figure 9a. For neat PGA, the compressive curve showed a sudden drop when the stain only reached 5% with a corresponding compression strength of 24.2 MPa, indicating an inherent brittleness of PGA. During the compression test, PGA D-surface structures broke into pieces when suffered compression loading force. In contrast, PGA/PBAT (85/15) structures displayed a continuous profile in the stress-strain curves, suggesting an improved toughness of composite materials. The maximum compressive strength was 29.9 MPa for PGA/PBAT (85/15), which was 25% higher than that of PGA. Accordingly, it was determined that PGA/PBAT had better resistance to compression loading force. Although the mechanical strength of neat PGA was stronger than PGA/PBAT blend, the brittleness constrained the applications when used as energy absorbers. The energy absorbed per mass of graded D-TPMS structure was calculated up to compressive strains to 0.2 and plotted in Figure 9b. The cumulative SEA continuously increased with the increase of compressive strain. After the compression test, samples exhibited a mixture failure mode of several delaminations and local cracks occurred at the bottom, which might be caused by minor stress fluctuations.

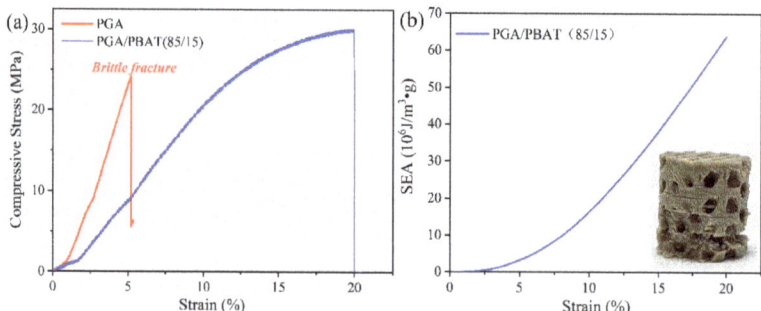

Figure 9. (a) Compressive stress-strain curves of PGA and PGA/PBAT (85/15) D-TPMS structures; (b) SEA versus strain curves of PGA/PBAT (85/15) structures.

Figure 10 displayed the FDM printed complicated parts using PGA/PBAT filament feedstock, proving that the PGA/PBAT blend filament can be 3D-printed into geometrically complex parts. In addition, vertically tall and slender cylinders were successfully printed with fine surfaces. Therefore, biodegradable PGA and PBAT polymer were suitable to manufacture high strength, lightweight and complex shaped parts, providing alternative materials of green plastics in the printing materials.

Figure 10. Typical FDM-3D printed PGA/PBAT industrial parts with complex shapes.

4. Conclusions

The research demonstrated that the binary PGA/PBAT (85/15) blend cross-linked by ADR chain extender was of good printability as FDM 3D printing feedstock. Utilizing biodegradable PGA/PBAT blends shows great potential in products or prototypes with a prospect of environmental protection value.

In conclusion:

(1) The crystallization process of composite filament was affected by blending of PBAT, and thermal stability of PGA/PBAT (95/5, 85/15) were superior to neat PGA whereas that of PGA/PBAT (75/25) became deteriorated. The utilization of an ADR chain extender can improve the compatibility of PGA and PBAT to some extent.

(2) The incorporation of PBAT decreased the tensile strength and modulus but effectively enhanced the elongation at the break of PGA/PBAT blends, achieving an improved toughness. The mechanical properties (including stiffness, toughness) could be well tailored by changing the formulations.

(3) 3D-printed PGA/PBAT (85/15) were successfully fabricated into filaments, and the mechanical performance of printed samples was close to that of injection-molded counterparts.

(4) D-TPMS structures with uniform and graded pore architectures were designed and manufactured. The graded-thickness PGA/PBAT TPMS samples exhibited good stiffness, strength and energy absorption capacities.

Future work could be focused on the development of wider compositions of PGA/PBAT blend filament for 3D printing and investigate the energy absorption between graded structures and non-graded structures.

Author Contributions: Methodology, B.W.; software, Y.Z. and F.H.; investigation, Z.Z.; data curation, Z.Z. and F.H.; writing—original draft preparation, L.S. and Z.W.; supervision, Y.Z.; project administration, L.S. and H.Z.; funding acquisition, H.Z. All authors have read and agreed to the published version of the manuscript.

Funding: This research was funded by "National Natural Science Foundation of China, grant number 52175216" and Dalian Medical Science Research Project (Nos. 2011001), "The APC was funded by Dalian Medical Science Research Project (Nos. 2011001)".

Institutional Review Board Statement: Not applicable.

Informed Consent Statement: Not applicable.

Data Availability Statement: The data presented in this study are available on request from the corresponding author. The data are not publicly available due to on-going relevant study.

Conflicts of Interest: The authors declare no conflict of interest.

References

1. Mohamed, O.A.; Masood, S.H.; Bhowmik, J.L. Optimization of fused deposition modeling process parameters: A review of current research and future prospects. *Adv. Manuf.* **2015**, *31*, 42–53. [CrossRef]
2. Ngo, T.D.; Kashani, A.; Imbalzano, G.; Nguyen, K.T.Q.; Hui, D. Additive manufacturing (3D printing): A review of materials, methods, applications and challenges. *Compos. Part B. Eng.* **2018**, *143*, 172–196. [CrossRef]
3. Wang, X.; Jiang, M.; Zhou, Z.; Gou, J.; Hui, D. 3D printing of polymer matrix composites: A review and prospective. *Compos. Part B. Eng.* **2017**, *110*, 442–458. [CrossRef]
4. Ligon, S.C.; Liska, R.; Stampfl, J.; Gurr, M.; Mulhaupt, R. Polymers for 3D Printing and Customized Additive Manufacturing. *Chem. Rev.* **2017**, *117*, 10212–10290. [CrossRef]
5. Ning, F.D.; Cong, W.L.; Qiu, J.J.; Wei, J.H.; Wang, S.R. Additive manufacturing of carbon fiber reinforced thermoplastic composites using fused deposition modeling. *Compos. Part B. Eng.* **2015**, *80*, 369–378. [CrossRef]
6. Turner, B.N.; Strong, R.; Gold, S.A. A review of melt extrusion additive manufacturing processes: I. Process design and modeling. *Rapid Prototyp. J.* **2014**, *20*, 192–204. [CrossRef]
7. Gremare, A.; Guduric, V.; Bareille, R.; Heroguez, V.; Latour, S.; L'Heureux, N.; Fricain, J.C.; Catros, S.; Le Nihouannen, D. Characterization of printed PLA scaffolds for bone tissue engineering. *J. Biomed. Mater. Res. Part A* **2018**, *106*, 887–894. [CrossRef]
8. Kuznetsov, V.E.; Solonin, A.N.; Urzhumtsev, O.D.; Schilling, R.; Tavitov, A.G. Strength of PLA Components Fabricated with Fused Deposition Technology Using a Desktop 3D Printer as a Function of Geometrical Parameters of the Process. *Polymers* **2018**, *10*, 313. [CrossRef] [PubMed]
9. Alizadeh-Osgouei, M.; Li, Y.C.; Wen, C.E. A comprehensive review of biodegradable synthetic polymer-ceramic composites and their manufacture for biomedical applications. *Bioact. Mater.* **2019**, *4*, 22–36. [CrossRef]
10. Ceylan, H.; Yasa, I.C.; Yasa, O.; Tabak, A.F.; Giltinan, J.; Sitti, M. 3D-Printed Biodegradable Microswimmer for Theranostic Cargo Delivery and Release. *ACS Nano* **2019**, *133*, 3353–3362. [CrossRef]
11. Hsieh, F.Y.; Hsu, S.H. 3D bioprinting: A new insight into the therapeutic strategy of neural tissue regeneration. *Organogenesis* **2015**, *11*, 153–158. [CrossRef] [PubMed]
12. Mazzanti, V.; Malagutti, L.; Mollica, F. FDM 3D Printing of Polymers Containing Natural Fillers: A Review of their Mechanical Properties. *Polymers* **2019**, *11*, 1094. [CrossRef]
13. Ji, J.Y.; Pang, Y.; Sutoko, S.; Horimoto, Y.; Sun, W.; Niino, T.; Sakai, Y. Design, Fabrication, and Evaluation of Polyglycolic Acid Modules with Canals as Tissue Elements in Cellular-Assembly Technology. *App. Sci.* **2020**, *10*, 3478.
14. Spearman, S.S.; Irin, F.; Rivero, I.V.; Green, M.J.; Abidi, N. Effect of dsDNA wrapped single-walled carbon nanotubes on the thermal and mechanical properties of polycaprolactone and polyglycolide fiber blend composites. *Polymers* **2015**, *56*, 476–481. [CrossRef]
15. Yeo, T.; Ko, Y.G.; Kim, E.J.; Kwon, O.K.; Chung, H.Y.; Kwon, O.H. Promoting bone regeneration by 3D-printed poly(glycolic acid)/hydroxyapatite composite scaffolds. *J. Ind. Eng. Chem.* **2021**, *94*, 343–351. [CrossRef]
16. Kim, B.N.; Ko, Y.G.; Yeo, T.; Kim, E.J.; Kwon, O.K.; Kwon, O.H. Guided Regeneration of Rabbit Calvarial Defects Using Silk Fibroin Nanofiber-Poly(glycolic acid) Hybrid Scaffolds. *Acs Biomater. Sci. Eng.* **2019**, *510*, 5266–5272. [CrossRef]
17. Santavirta, S.; Konttinen, Y.T.; Saito, T.; Gronblad, M.; Partio, E.; Kemppinen, P.; Rokkanen, P. Immune response to polyglycolic acid implants. *J. Bone Jt. Surg. Br.* **1990**, *72*, 597–600. [CrossRef] [PubMed]
18. Wu, P.; Hu, S.; Liang, Q.; Guo, W.; Xia, Y.; Shuai, C.; Li, Y. A polymer scaffold with drug-sustained release and antibacterial activity. *Int. J. Polym. Mater. Polym. Biomater.* **2020**, *69*, 398–405. [CrossRef]
19. Choi, J.; Hong, G.; Kwon, T.; Lim, J.O. Fabrication of Oxygen Releasing Scaffold by Embedding H_2O_2-PLGA Microspheres into Alginate-Based Hydrogel Sponge and Its Application for Wound Healing. *Appl. Sci.* **2018**, *8*, 1492. [CrossRef]
20. Jiao, J.; Zeng, X.; Huang, X. An overview on synthesis, properties and applications of poly(butylene-adipate-co-terephthalate)-PBAT. *Adv. Ind. Eng. Polym. Res.* **2020**, *31*, 19–26.
21. Di Lorenzo, M.L. Poly(l-Lactic Acid)/Poly(Butylene Succinate) Biobased Biodegradable Blends. *Polym. Rev.* **2021**, *61*, 457–492. [CrossRef]
22. Hivechi, A.; Bahrami, S.H.; Siegel, R.A. Drug release and biodegradability of electrospun cellulose nanocrystal reinforced polycaprolactone. *Mater. Sci. Eng. C* **2019**, *94*, 929–937. [CrossRef] [PubMed]
23. Shuai, C.; Wu, P.; Zhong, Y.; Feng, P.; Gao, C.; Huang, W.; Zhou, Z.; Chen, L.; Shuai, C. Polyetheretherketone/poly (glycolic acid) blend scaffolds with biodegradable properties. *J. Biomater. Sci. Poly. Ed.* **2016**, *27*, 1434–1446. [CrossRef] [PubMed]
24. Vartiainen, J.; Shen, Y.F.; Kaljunen, T.; Malm, T.; Vaha-Nissi, M.; Putkonen, M.; Harlin, A. Bio-based multilayer barrier films by extrusion, dispersion coating and atomic layer deposition. *J. App. Poly. Sci.* **2016**, *133*, 42260. [CrossRef]
25. Zhu, Y.; Wang, Z.; Zhou, H.; Li, L.; Zhu, Q.; Zhang, P. An injectable hydroxyapatite/poly(lactide-co-glycolide) composite reinforced by micro/nano-hybrid poly(glycolide) fibers for bone repair. *Mater. Sci. Eng. C* **2017**, *80*, 326–334. [CrossRef]
26. Lyu, Y.; Chen, Y.; Lin, Z.; Zhang, J.; Shi, X. Manipulating phase structure of biodegradable PLA/PBAT system: Effects on dynamic rheological responses and 3D printing. *Compos. Sci. Technol.* **2020**, *200*, 100839. [CrossRef]

27. Prasong, W.; Muanchan, P.; Ishigami, A.; Thumsorn, S.; Kurose, T.; Ito, H. Properties of 3D Printable Poly(lactic acid)/Poly(butylene adipate-co-terephthalate) Blends and Nano Talc Composites. *J. Nanomater.* **2020**, *2020*, 8040517. [CrossRef]
28. Shen, J.; Wang, K.; Ma, Z.; Xu, N.; Pang, S.; Pan, L. Biodegradable blends of poly(butylene adipate-co-terephthalate) and polyglycolic acid with enhanced mechanical, rheological and barrier performances. *J. Appl. Polym. Sci.* **2021**, *138*, 51285. [CrossRef]
29. Xue, P.; Xu, F.; Xu, L. Epoxy-functionalized mesostructured cellular foams as effective support for covalent immobilization of penicillin G acylase. *Appl. Surf. Sci.* **2008**, *2555*, 1625–1630. [CrossRef]
30. Al-Ketan, O.; Abu Al-Rub, R.K.; Rowshan, R. Mechanical Properties of a New Type of Architected Interpenetrating Phase Composite Materials. *Adv. Mater. Technol.* **2017**, *2*, 1600235. [CrossRef]
31. Jin, Y.; Kong, H.; Zhou, X.; Li, G.; Du, J. Design and Characterization of Sheet-Based Gyroid Porous Structures with Bioinspired Functional Gradients. *Materials* **2020**, *13*, 3844. [CrossRef]
32. Ashby, M.F. The mechanical properties of cellular solids. *Metall. Trans. A* **1983**, *14*, 1755–1769. [CrossRef]
33. Zhang, L.; Feih, S.; Daynes, S.; Chang, S.; Wang, M.; Wei, J.; Lu, W. Energy absorption characteristics of metallic triply periodic minimal surface sheet structures under compressive loading. *Addit. Manuf.* **2018**, *23*, 505–515. [CrossRef]
34. Huang, D.; Hu, Z.; Liu, T.; Lu, B.; Zhen, Z.; Wang, G.; Ji, J. Seawater degradation of PLA accelerated by water-soluble PVA. *E-Polymers* **2020**, *20*, 759–772. [CrossRef]
35. Tang, D.; Zhang, C.; Weng, Y. Effect of multi-functional epoxy chain extender on the weathering resistance performance of Poly(butylene adipate-co-terephthalate) (PBAT). *Polym. Test.* **2021**, *99*, 107204. [CrossRef]
36. Zehetmeyer, G.; Meira, S.M.M.; Scheibel, J.M.; de Oliveira, R.V.B.; Brandelli, A.; Soares, R.M.D. Influence of melt processing on biodegradable nisin-PBAT films intended for active food packaging applications. *J. Appl. Polym. Sci.* **2016**, *133*, 43212. [CrossRef]
37. Yu, C.; Bao, J.; Xie, Q.; Shan, G.; Bao, Y.; Pan, P. Crystallization behavior and crystalline structural changes of poly(glycolic acid) investigated via temperature-variable WAXD and FTIR analysis. *Crystengcomm* **2016**, *18*, 7894–7902. [CrossRef]
38. Li, W.; Sang, L.; Jian, X.; Wang, J. Influence of sanding and plasma treatment on shear bond strength of 3D-printed PEI, PEEK and PEEK/CF. *Int. J. Adhes. Adhes.* **2020**, *100*, 102614. [CrossRef]
39. Chacon, J.M.; Caminero, M.A.; Nunez, P.J.; Garcia-Plaza, E.; Garcia-Moreno, I.; Reverte, J.M. Additive manufacturing of continuous fibre reinforced thermoplastic composites using fused deposition modelling: Effect of process parameters on mechanical properties. *Compos. Sci. Technol.* **2019**, *181*, 107688. [CrossRef]
40. Prasong, W.; Ishigami, A.; Thumsorn, S.; Kurose, T.; Ito, H. Improvement of Interlayer Adhesion and Heat Resistance of Biodegradable Ternary Blend Composite 3D Printing. *Polymers* **2021**, *13*, 740. [CrossRef]
41. Liu, F.; Mao, Z.; Zhang, P.; Zhang, D.; Jiang, J.; Ma, Z. Functionally graded porous scaffolds in multiple patterns: New design method, physical and mechanical properties. *Mater. Des.* **2018**, *160*, 849–860. [CrossRef]
42. Zhang, X.; Fang, G.; Xing, L.; Liu, W.; Zhou, J. Effect of porosity variation strategy on the performance of functionally graded Ti-6Al-4V scaffolds for bone tissue engineering. *Mater. Des.* **2018**, *157*, 523–538. [CrossRef]
43. Abou-Ali, A.M.; Al-Ketan, O.; Rowshan, R.; Abu Al-Rub, R. Mechanical Response of 3D Printed Bending-Dominated Ligament-Based Triply Periodic Cellular Polymeric Solids. *J. Mater. Eng. Perform.* **2019**, *284*, 2316–2326. [CrossRef]

Article

Comparison between Tests and Simulations Regarding Bending Resistance of 3D Printed PLA Structures

Dorin-Ioan Catana [1,*], Mihai-Alin Pop [2] and Denisa-Iulia Brus [3]

1 Department of Materials Engineering and Welding, Transilvania University of Brasov, 500036 Brasov, Romania
2 Department of Materials Science, Transilvania University of Brasov, 500036 Brasov, Romania; mihai.pop@unitbv.ro
3 Department of Motor Performance, Transilvania University of Brasov, 500036 Brasov, Romania; denisa.brus@unitbv.ro
* Correspondence: catana.dorin@unitbv.ro

Abstract: Additive manufacturing is one of the technologies that is beginning to be used in new fields of parts production, but it is also a technology that is constantly evolving, due to the advances made by researchers and printing equipment. The paper presents how, by using the simulation process, the geometry of the 3D printed structures from PLA and PLA-Glass was optimized at the bending stress. The optimization aimed to reduce the consumption of filament (material) simultaneously with an increase in the bending resistance. In addition, this paper demonstrates that the simulation process can only be applied with good results to 3D printed structures when their mechanical properties are known. The inconsistency of printing process parameters makes the 3D printed structures not homogeneous and, consequently, the occurrence of errors between the test results and those of simulations become natural and acceptable. The mechanical properties depend on the values of the printing process parameters and the printing equipment because, in the case of 3D printing, it is necessary for each combination of parameters to determine their mechanical properties through specific tests.

Keywords: additive manufacturing; poly(lactic acid); optimization; simulation; finite element analysis (FEA)

Citation: Catana, D.-I.; Pop, M.-A.; Brus, D.-I. Comparison between Tests and Simulations Regarding Bending Resistance of 3D Printed PLA Structures. *Polymers* **2021**, *13*, 4371. https://doi.org/10.3390/polym13244371

Academic Editors: Swee Leong Sing and Wai Yee Yeong

Received: 9 November 2021
Accepted: 10 December 2021
Published: 14 December 2021

Publisher's Note: MDPI stays neutral with regard to jurisdictional claims in published maps and institutional affiliations.

Copyright: © 2021 by the authors. Licensee MDPI, Basel, Switzerland. This article is an open access article distributed under the terms and conditions of the Creative Commons Attribution (CC BY) license (https://creativecommons.org/licenses/by/4.0/).

1. Introduction

The twentieth century was marked by the unprecedented development of the engineering sciences. This evolution was possible due to the important steps made in the theoretical field by other disciplines, including mathematics, physics, and chemistry. The demonstration of some basic theories and theorems in mathematics and physics, as well as the inclusion of mathematical analysis and differential equations in the solution of engineering problems, meant finding theoretical solutions to practical aspects encountered in engineering. The end of the previous century allowed the development of technologies and equipment for 3D printing. Additive processing is a process of continuous exploitation and exploration. Exploitation because new and new applications in different fields are solved using this technology, and exploration because researchers attempt to find solutions to the hitherto unresolved aspects of this technology, which, as research shows are not few.

Additive manufacturing has seen the rapid expansion and continuous development of printable materials. 3D printing is currently widespread and it has begun to address new areas through its involvement in different forms of goods manufacturing and in sports (such as alpine skiing). Given the relatively short time in which that 3D printing has been used, this technology still features many aspects that need to be understood, solved, and improved. Currently, producers want to reduce the time it takes for products to reach the market and for consumers (beneficiaries) to purchase them at the lowest possible cost. The way to improve the properties of 3D printed structures is through the constant attention of researchers, who attempt to apply different techniques to reach the proposed objectives. This approach consists in applying the capabilities of the simulation process to 3D printed parts, to optimize the respective products. Bibliographic study demonstrates that currently, there are few works that address the implementation of the simulation (modelling) process, with its undeniable benefits, in additive manufacturing.

The printing parameters can significantly influence the mechanical properties of the 3D printing parts (printing speed and nozzle temperatures). Furthermore, with a finite element analysis (FEA) the stress distribution of single-tensile testing, bending testing, and compression testing of poly(lactic acid)-PLA samples has been visualized [1]. The simulation can provide critical inputs for the designer. Moreover, based on experimental data obtained (extracted) from previous research, a finite element analysis can be applied. Studies reveal that the deviations between simulation and experimental results were minimal, and the maximum error was 6.7%. In this way, the simulation could be used to predict the behaviour of 3D printed parts [2]. Experimental and theoretical results demonstrate that the tensile strength of 3D printed poly(lactic acid) decrease as the layer thickness increases from 0.1 to 0.3 mm. Furthermore, the experimental results demonstrate that the ultimate tensile strength of 3D printed samples changes significantly with changes in the printing angle [3]. Simulation results show that strain and displacement in the gage region offer results that are comparable with experimental results [4]. The improvement of the 3D printed part quality requires many studies about the optimal setting for additive manufacturing parameters [5]. Research shows that 3D printing parameters significantly influence the elastic strength of polymer composites [6]. The best results from the tests and simulation were obtained when the infill pattern was 100%. In addition, the difference between the experimental results and the simulation was below 10% [7,8]. With a low filament consumption, the researchers showed that the values of the printing parameters can be optimized in the case of additive manufacturing, based on material extrusion [9]. The simulation can be applied to predict the behaviour of the 3D printed structures, from PLA [10].

The study of the published articles on the additive manufacturing of poly(lactic acid) led to the following conclusions:

- material is assimilated as homogenous and linear-elastic [11];
- anisotropy of 3D printed structures is mild [12];
- Poisson's ratio for the 3D printed parts is between 0.33–0.36.

All this information highlights that there are necessary bases for the application of the simulation process for the PLA structures, obtained by additive manufacturing. Furthermore, Refs. [13,14] demonstrate that the simulation process can be applied with good results to the 3D printed structures from PLA, because the differences between the test results and those of the simulation feature reduced errors. The simulations were performed based on the results obtained from the tensile and bending tests performed on the 3D printed structures from PLA. In previous studies [13,14], the mechanical properties of the 3D printed structures were determined from the used filaments, because there were significant differences between the properties of the filaments and those of the printed structures. It should be noted that 3D printing involves the filament melting, followed by its solidification. Thus, the PLA obtained structures by additive manufacturing consist in solidifications of the deposited filament, in successive layers. This approach to generating 3D printed structures influences their resistance to different stresses (loadings). The explanation is that between the successively deposited layers (one already solidified and the other in the process of solidification), strains appear, which can reduce, to a greater or lesser extent, the properties established by the calculus. The reason why 3D printed structures need to be tested is to determine their mechanical properties. When the mechanical properties are determined, they can be passed to the optimization stage by simulation.

Poly(lactic acid) was chosen because it is one of the most commonly used filaments in additive manufacturing, and it also features many applications in the medical field.

In addition, the tensile and bending stresses were studied because these are the stresses that develop most often in the 3D printed structures from PLA. The aim of the paper is to demonstrate that, based on existing or determined information regarding the physical and mechanical properties of 3D printing, finite element analysis (FEA or FEM-finite element modelling) can be applied to optimize the respective structure from the geometric and dimensional point of view. The results of the simulation process, which are in line with those of theory and previous tests, make it possible to reduce the design time, while improving the behaviour at various stress points for 3D printed structures.

2. Experimental Setup

The materials (filaments) used in the specimen printing were poly(lactic acid) (PLA white manufactured by Suntem3D, Bucharest, Romania) and poly(lactic acid) mixed up to 20% with glass fibre (PLA-Glass, manufactured by Filaticum, Miskolc, Hungary). For the PLA, the mechanical properties included: filament diameter 2.85 mm, tensile strength 1100 MPa (ASTM D882), modulus of elasticity 3310 MPa (AST MD882), and bending modulus of elasticity 2392.5 MPa. For the PLA-Glass, the mechanical properties included: filament diameter 2.85 mm, maximum tensile strength 57 MPa (ASTM D638), tensile strength at yield 46 MPa (ASTM D638), tensile modulus 4.0 GPA (ASTM D638), and tensile elongation 3.4% (ASTM D638). The complete technical characteristics of both filaments are presented in their respective technical data sheets.

From the mentioned filaments, the 3D specimens to be used in the bending tests were printed, in order to optimize them from a geometric and dimensional point of view. For the 3D printing parameters with which the PLA structures were processed, the mechanical properties were determined and presented in previous works [13,14]. In the case of additive manufactured structures, for the simulation process to be performed, the mechanical properties necessary are the following:

- density of the 3D printed specimen; this is not equal with the density of the filament, because it depends on the 3D structure's printed parameters;
- bending deflection;
- yield strength;
- modulus of elasticity;
- ultimate strength at bending (bending strength);
- Poisson's ratio.

The printing of the specimens performed on a CreatBot DX-3D double-nozzle printer (manufacturer Henan Suwei Electronic Technology Co., LTD., Zhengzhou, China). The printer capabilities were:

- printing dimensions—300 × 250 × 300 mm;
- filament diameter—2.8–3.0 mm;
- printing nozzle—0.2–0.8 mm;
- printing resolution—0.6 mm;
- layer resolution—0.2 mm;
- printing volume—22.5 l.

The parameters of the printing process were:

- layer height—0.2 mm;
- printing temperature—210 °C;
- print speed—50 mm/s;
- printing angle (overhang angle for support)—45°;
- bed temperature—61 °C;
- infill—100% (the internal structure is solid; with a solid infill at the top and bottom);
- infill overlap—10%;
- infill flow—110%.

The printed structures from the PLA are of bar or tube type, with circular, elliptical or rectangular sections. To verify the efficiency of the simulation process, specimens with a profile I section were printed. These specimens were bar type. Furthermore, tube type specimens with rectangular sections, but consolidated in the middle, were produced (cross-section Rect-cons). Additive manufacturing specimens were obtained from the PLA filament (100% PLA) or PLA-Glass filament (100% PLA-Glass). Table 1 presents the coding of the specimens according to the material, the geometry of the section, the dimensions, and the type.

Table 1. Specimen characteristics and codification (coding).

Filament Type	Cross Section	Dimensions (mm)	Specimen Type	Specimen Code
PLA	Circular	12	Bar	P_12
PLA-Glass	Circular	12	Bar	G_12
PLA	Circular	12×10	Tube	P_12_T
PLA-Glass	Circular	12×10	Tube	G_12_T
PLA	Ellipse	18×8	Bar	P_E18
PLA-Glass	Ellipse	18×8	Bar	G_E18
PLA	Ellipse	18×16	Tube	P_E18_T
PLA-Glass	Ellipse	18×16	Tube	G_E18_T
PLA	Rectangular	12.6×9	Bar	P_R12
PLA-Glass	Rectangular	12.6×9	Bar	G_R12
PLA	Rectangular	12.6×10.8	Tube	P_R12_T
PLA-Glass	Rectangular	12.6×10.8	Tube	G_R12_T
PLA	I-section	11.2×28	Bar	P_IS
PLA-Glass	I-section	11.2×28	Bar	G_IS
PLA	Rect-cons	12.6×10.8	Tube	P_RC_T
PLA-Glass	Rect-cons	12.6×10.8	Tube	G_RC_T

The shapes and dimensions of the specimens are presented in Figures 1 and 2.

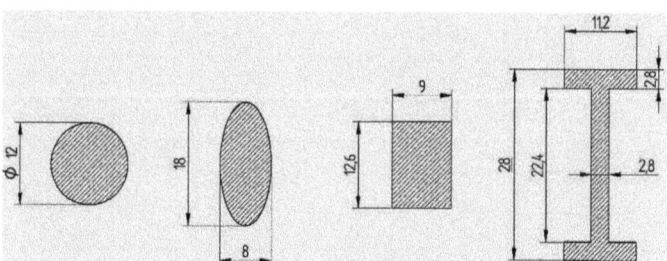

Figure 1. Shapes and dimensions of the bar-type specimens (ISO metric drawing standard).

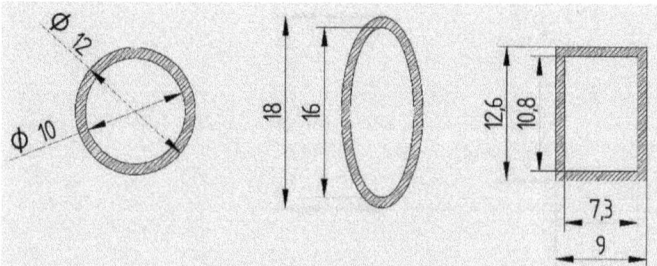

Figure 2. Shapes and dimensions of the tube-type specimens (ISO metric drawing standard).

The 3D printed specimens were tested at bending and the equipment used was a WDW-150S Universal testing machine. The bending tests were performed under the following conditions:

- bending (loading) speed 10 mm/min;
- stress speed 10 MPa/s;
- support—cylindrical (diameter 30 mm, length 70 mm);
- loading nose/anvil—semi-cylindrical (diameter 30 mm, length 70 mm).

On this testing machine, the test force can be modified between 0.1 and 150 kN.

Due to the bending stress, the straight beams (girder) become deformed and curve-shaped until they exceed a critical value, at which point they break. Bending tests allow the bending strength and deformation (displacement) to be obtained. Deformation, also known as the arrow, represents the maximum transverse displacement produced in the middle of the opening of a beam supported at its ends. The value of the deformation represents the shape that the beam can form when it is bent under stress or the deformations produced near some sections. The bending tests made it possible to determinate the maximum force that produced the breaking (rupture) of the specimens and, based on this, it was possible to calculate the bending strength. The tests were performed for both the printed PLA and the PLA-Glass specimens. The tests were also performed for the bar- or tube-type specimens and for the geometries displayed in Table 1. By complementing the information established during other tests and available in various papers with those obtained during the tests presented in the paper, the simulation process can be started.

Before presenting and discussing the results of the simulation, it should be noted that regardless of the geometry of the cross-section, the value of the section is the same for the bar specimens (113 mm^2). The tube type specimens also feature the same value for the respective section, but of course, the value is significantly reduced (34.5 mm^2, which is 31% of the bar type section's value). In this way, the efficiency of each type of specimen and geometry can be better understood, while comparisons between results can be made more easily. The bending force was applied halfway between the supports, where the deformation of the specimen was maximal. Based on the test results, the simulation process could be applied, which was performed under the same conditions as the tests, which were as follows:

- the stress (load) force was equal to the one during the tests;
- the force was applied halfway between the supports;
- the distance between the supports was 180 mm;
- the physical and mechanical properties of the 3D printed structures were those established by the tests, previously presented.

For the geometry of the section, we opted for the shapes described in Figures 1 and 2. According to the theory, the bending strength depends on the moment of inertia (I_z) and the distance between the point on the surface for which the calculus is made and the neutral axis (y_{max}). The ratio between I_z and y_{max} is called the axial resistance moment (W_z) and is a geometric feature of the cross section. The bending strength for the described stress scheme is calculated by the relation:

$$\sigma_{max} = \frac{M_i \cdot y_{max}}{I_z} = \frac{M_i}{W_z} = \frac{F_{max} \cdot l}{4 \cdot W_z}, \qquad (1)$$

where M_i is the bending moment, W_z the axial resistance moment (modulus), F_{max} the maximum force that produced the rupture of the specimen, and l is the distance between the supports. The study of relation (1) shows that in order to obtain the lowest possible bending resistance, the geometry of the structure must feature an axial resistance modulus that is as large as possible because it is assumed that the distance between the supports is kept constant. In addition to the bar-type specimens, tube-type specimens with the same cross-sectional geometries as those of the filled section (bar) specimens were printed.

The finite element analysis was performed with the Simulation module in Solid Edge ST10 2D and 3D software for engineers (Siemens Industry Software Inc., Plano, TX, USA). The simulation process was performed under the following conditions: mesh type–

tetrahedral; study–linear static; meshing level–9 for all simulations, which generated a mesh size between 1.6 and 3.45 mm (depending on specimens' type and their geometries).

3. Results and Discussion

The specimens obtained through additive manufacturing were tested for bending. As mentioned, this stress (load) is common in 3D printed structures. The simulation results are presented in Table 2.

Table 2. Comparison between results of the tests and simulation, for 3D printed specimens, bending-stressed.

Specimen Cod	Test Results		Simulation Results	
	Strength (MPa)	Deformation (mm)	Strength (MPa)	Deformation (mm)
P_12	81.70	15.20	81.80	14.40
G_12	26.50	6.20	27.60	4.10
P_12_T	40.10	7.50	39.40	5.90
G_12_T	28.90	7.80	30.40	3.80
P_E18	54.90	5.80	55.30	6.10
G_E18	41.40	4.10	42.50	4.50
P_E18_T	82.30	6.40	83.10	8.20
G_E18_T	36.10	3.40	36.50	3.20
P_R12	62.70	9.10	61.50	9.70
G_R12	29.10	6.00	29.30	4.80
P_R12_T	35.50	4.60	38.60	5.20
G_R12_T	26.60	6.50	28.90	4.80
P_IS	45.70	4.20	47.80	4.90
G_IS	20.80	2.60	22.30	2.90
P_RC_T	32.10	5.70	33.80	6.80
G_RC_T	19.70	2.60	20.10	3.70

Figure 3 depicts the simulation results for the specimen with the rectangular section obtained from the PLA filament.

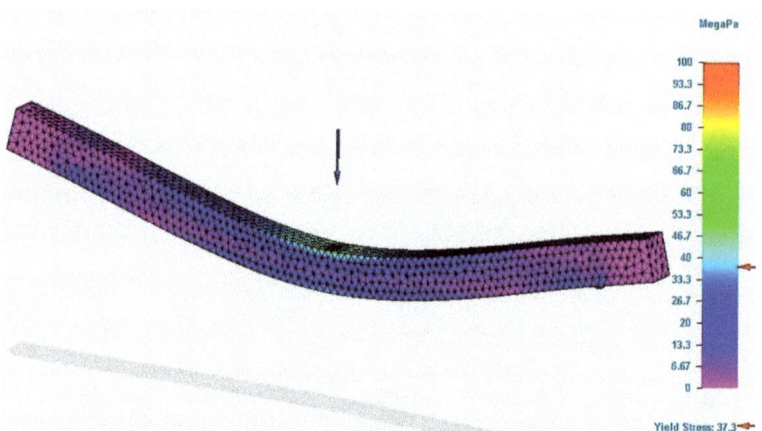

Figure 3. Results of the simulation process for P_R12 specimen (Von Mises stress).

The figure presents the value of the Von Mises stresses developed in the specimen, when a force equal to that obtained in the bending tests was applied. In the force application area, the figure demonstrates that the value of the bending stress was close to that determined by tests. The value determined by the simulation was smaller than the one registered in the test. Regarding the deformation value (see Figure 4), it was found that it was greater than the real value, namely, the value recorded by the test equipment.

Regarding the deformation, it should be mentioned that the theory demonstrates that the analytical method of integration of the differential equation is applied for an approximate deformed average fiber. Therefore, there are possible errors in the calculus compared to reality.

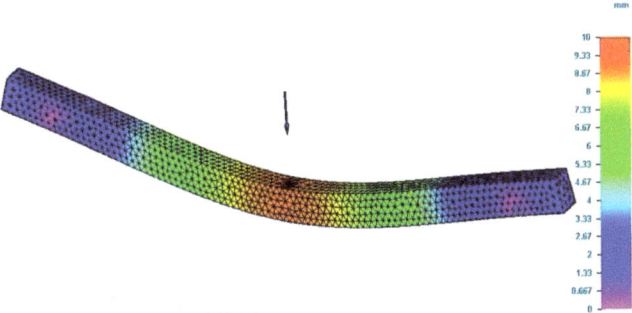

Figure 4. Results of simulation process for P_R12 specimen (total deformation values).

In Figure 5, the simulation results for the tube-type specimen with rectangular section obtained from PLA filament is presented.

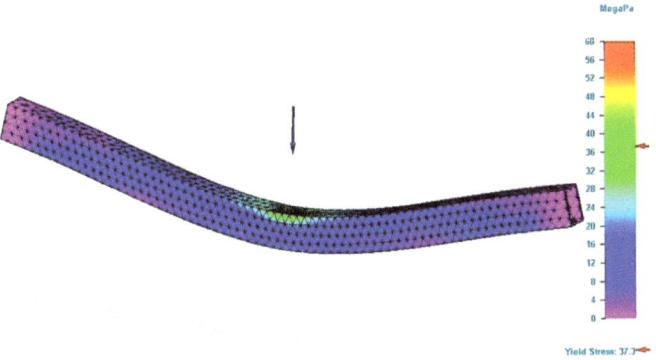

Figure 5. Results of simulation process for P_R12_T specimen (Von Mises stress).

The simulation of the deformation for the specimen P_R12_T is presented in Figure 6. Using the capabilities of the simulation program, the value of the maximum deformation that occurs in the specimen before it ruptures can be determined.

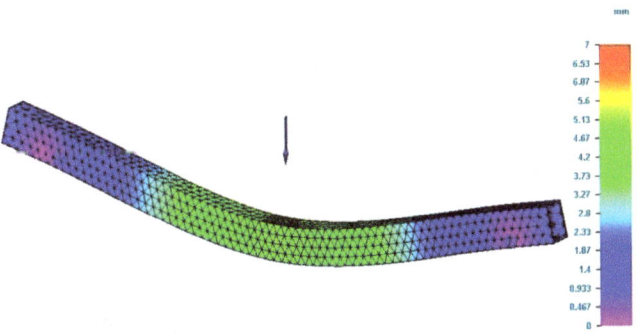

Figure 6. Results of simulation process for P_R12_T specimen (total deformation values).

The value obtained by the simulation was greater than the value recorded during the test. This error can be explained by the approximate methods of calculating the average fibre, but also by the elasticity of the tube-type specimen.

Analyzing Table 2, it can be seen that the application of the simulation provided results close to those of the tests. In addition, the test and simulation data demonstrate that it was possible to optimize the geometry of the 3D printed structures. This statement is based on the following aspects. As mentioned, the bar-type specimens featured equal sections, as did the tube-type specimens. The length was 220 mm for all the 3D printed specimens. The deduction is that the volume was the same ($V = A \cdot l$) for the bar or tube-type specimens, including their mass. All this occurred because it depended on the volume and density ($m = V\rho$). The density of the specimens remained constant if the printing parameters were not changed.

Therefore, the efficiency of the section geometry is given by the ratio between the axial resistance modulus and the surface. When the W_z is higher, for the same surface of the 3D printed structure, the specimen can be stressed with higher forces until breaking. Thus, the ratio between the axial resistance modulus and the area can be considered an indicator of the efficiency of the section geometry, or it can reveal whether the section is optimal. Therefore, as this ratio is greater, the volume of the printed material in this structure is smaller, but with the same bending strength, or even higher.

For the validation of the presented features, structures with section I were printed by additive manufacturing. The respective structure featured a cross section 11% larger than the bar-type structures used in the study. The dimensions of the surface made it possible to increase the height of the specimen, which had a significant impact on the ratio between the breaking force and the volume (see the last column in Table 3).

Table 3. Efficiency of the specimen cross-section.

Specimen Code	Cross Section	Surface (mm^2)	Volume (mm^3)	W_z (mm^3)	W_z/S (mm)	F/V (N/mm^3)
P_12	Circular	113	24,900	170	1.50	0.012
G_12	Circular	113	24,900	170	1.50	0.004
P_E18	Ellipse	113	24,900	254	2.20	0.013
G_E18	Ellipse	113	24,900	254	2.20	0.009
P_R12	Rectangular	113	24,900	238	2.10	0.013
G_R12	Rectangular	113	24,900	238	2.10	0.006
P_12_T	Circular	34.50	7600	88	2.50	0.010
G_12_T	Circular	34.50	7600	88	2.50	0.007
P_E18_T	Ellipse	34.50	7600	115	3.30	0.028
G_E18_T	Ellipse	34.50	7600	115	3.30	0.012
P_R12_T	Rectangular	34.50	7600	116	3.40	0.012
G_R12_T	Rectangular	34.50	7600	116	3.40	0.009
P_IS	I-section	125	27,600	902	7.20	0.032
G_IS	I-section	125	27,600	902	7.20	0.015
P_RC_T	Rect-cons	71.70	8100	174	2.40	0.015
G_RC_T	Rect-cons	71.70	8100	174	2.40	0.009

Furthermore, in the case of the rectangular specimens with the tube section, a consolidation was performed in the middle of the specimen (see Figure 7). The efficiency of this consolidation can be followed by studying the values in the last column of Table 3. The comment that needs to be made is that the value of the section is valid only for the middle of the specimen, where it decreases by half, reaching the value of the tube-type sections. The application of the consolidation demonstrates that the force that caused the specimen rupture was greater than the force at which the specimens broke without this modification.

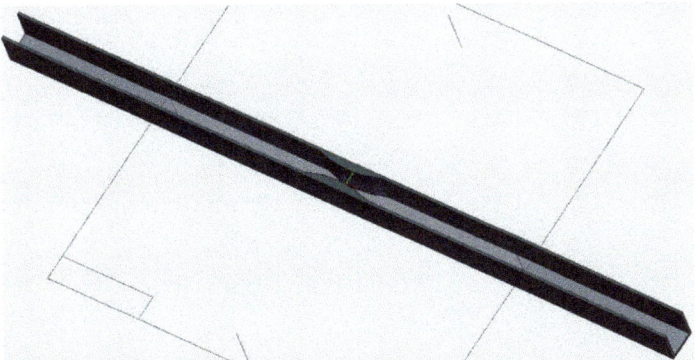

Figure 7. Rectangular specimen with consolidation (inside view).

The positive evolution is valid for both materials used in the study. From the point of view of volume, its increase by 6.5% for the reinforced rectangular specimens determined an increase in the breaking force by 34.80% for the P_RC_T specimens and by 10.4% for the G_RC_T specimens. In other words, with a small addition of material in well-defined places, the bending strength increases (see Figure 8).

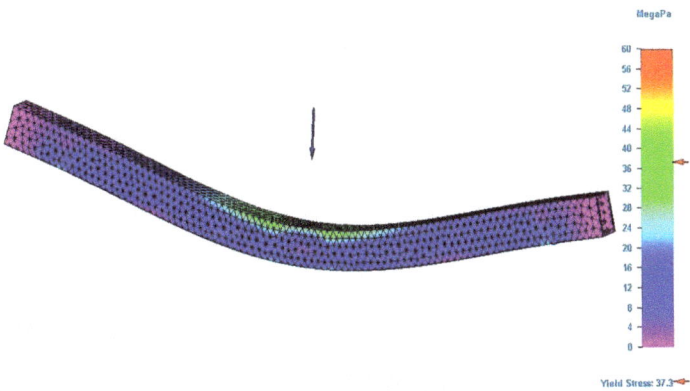

Figure 8. Results of simulation process for P_RC_T specimen (Von Mises stress).

Analyzing Figures 5 and 8, the differences between the stresses generated in the same zone of the specimen are clearly demonstrated. This evolution is favorable for the specimens that benefited from consolidation. Thus, the comparison demonstrates that the simulation process can be applied with confidence in the additive manufacturing of PLA structures and can be considered a useful and important tool for designers. The presented features demonstrate that by optimizing the geometries of the additive manufactured structures using the simulation process, it is possible to substantially improve their behavior at the stresses that are applied to them. A smaller volume of printed filament also means lower energy consumption for the structure processing. Implicitly, at the end of the lifecycle of the structure, the volume to be recycled is lower.

The difference between the test results and those of the simulation for bending strength was between—2.0% and 7.50% for the bar-type specimens and between—1.80% and 8.80% for the tube-type specimens. For the deformations, the differences were between −32.80% and 16.30% for the bar-type specimens and between −4.50% and 51.50% for the tube-type specimens. In the deformation, the value of the errors was higher; this was firstly due to the approximations of the deformed average fibre and, secondly, to the higher elasticity of the

3D printed structures of the tube-type. Furthermore, some deformation values were very small (2.60 mm), which may have led too error. For an easier and correct understanding of the geometrical efficiency, the results from Table 3 (last column) are presented in graphical form (see Figure 9).

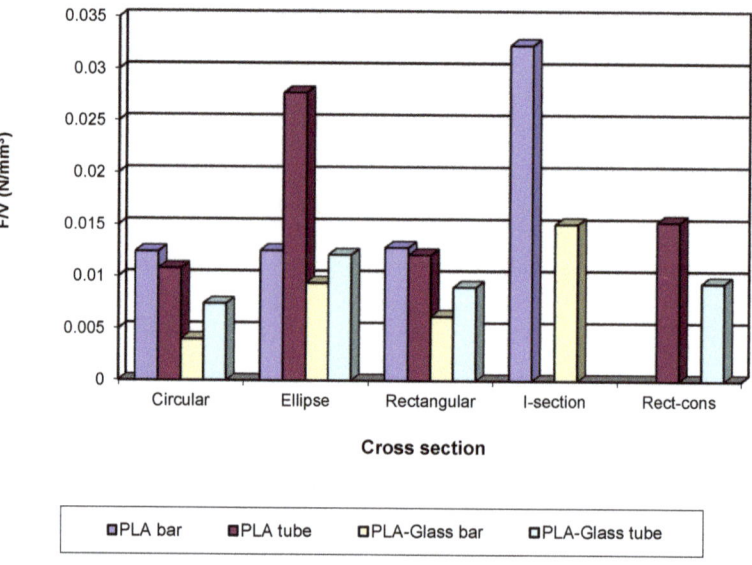

Figure 9. Efficiency of section geometry.

4. Conclusions

Three-dimensional printing is a relatively new manufacturing process compared to the processes used by humans since antiquity, or even earlier. The novelty of the process did not prevent it from being implemented in many fields of goods production. The optimization of structures processed by different methods has become a necessity as the engineering sciences have developed, and computational technology has increased its computing capacity and speed. Through optimization, the material consumption decreases, while the geometry of the processed structures receives a shape that makes it possible to increase their loading capacity.

The presented study demonstrates that by applying the simulation process, it is possible to optimize the geometry of the section for 3D printed structures. To check whether the optimization by simulation of the geometry can be applied or not, several geometries of the section, several types of structures (bar, tube) and two materials were chosen. In this study, it was demonstrated that the use of simulation in the optimization process leads to obtaining results in line with those determined by tests.

This study demonstrates that the simulation process provides results close to those of tests and in line with the results presented in previously published. In some cases, the values of the deformations evolved inappropriately evolution, but upon analysis, it was possible to identify the causes. Consequently, by improving the modeling process, the number of errors can be diminished. For the bar-type specimens, from the bending resistance point of view, the most advantageous section was profile I and, after this, the elliptical section. For the tube-type specimens, the most advantageous section was the elliptical, followed by the rectangular section with consolidation. As mentioned, the results of the simulation process demonstrate similar results for the bending strength to those of the tests; however, in the area of force application, there was an increase in deviations compared to the values obtained through the tests. A possible solution is the restoration of the simulations but using a digital replica the elements that interact during the tests that is as accurate as possible. More specifically, this step means moving from a schematic to a more complex representation of the elements involved in the bending test (support and loading nose/anvil).

When the mechanical properties of 3D printed structures are known, the simulation process can be applied with good results in order to optimize the geometry of those structures. Depending on the progress that is registered in the field of simulation programs, as well as in the theory of the materials' strength, the differences highlighted in this paper will possibly be reduced. Furthermore, other studies may yield new information on mechanical and technological properties, depending on the printing parameters, which will help to improve the simulation process; more precisely the differences between the simulation results and those of the tests will be reduced.

Author Contributions: Conceptualization, D.-I.C.; methodology, D.-I.C.; software, D.-I.C. and D.-I.B.; validation, D.-I.C., M.-A.P. and D.-I.B.; formal analysis, D.-I.C.; investigation, D.-I.C. and M.-A.P.; resources, D.-I.C. and M.-A.P.; data curation, D.-I.C. and M.-A.P.; writing—original draft preparation, D.-I.C. and D.-I.B.; writing—review and editing, D.-I.C. and D.-I.B.; visualization, D.-I.C., M.-A.P. and D.-I.B.; supervision, D.-I.C., M.-A.P. and D.-I.B.; project administration, D.-I.C.; funding acquisition, M.-A.P. All authors have read and agreed to the published version of the manuscript.

Funding: This research was funded by Transilvania University of Brasov, Brasov, Romania.

Institutional Review Board Statement: Not applicable.

Informed Consent Statement: Not applicable.

Conflicts of Interest: The authors declare no conflict of interest.

References

1. Abeykoon, C.; Sri-Amphorn, P.; Fernando, A. Optimization of fused deposition modeling parameters for improved PLA and ABS 3D printed structures. *Int. J. Lightweight Mater. Manuf.* **2020**, *3*, 284–297. [CrossRef]
2. Alharbi, M.; Kong, I.; Patel, V.I. Simulation of uniaxial stress–strain response of 3D-printed polylactic acid by nonlinear finite element analysis. *Appl. Adhes. Sci.* **2020**, *8*, 1–10. [CrossRef]
3. Yao, T.; Deng, Z.; Zhang, K.; Li, S. A method to predict the ultimate tensile strength of 3D printing polylactic acid (PLA) materials with different printing orientations. *Compos. Part B Eng.* **2019**, *163*, 393–402. [CrossRef]
4. Alafaghani, A.; Qattawi, A.; Alrawi, B.; Guzman, A. Experimental optimization of fused deposition modelling processing parameters: A design-for-manufacturing approach. *Procedia Manuf.* **2017**, *10*, 791–803. [CrossRef]
5. Chacón, J.M.; Caminero, M.A.; García-Plaza, E.; Núñez, P.J. Additive manufacturing of PLA structures using fused deposition modelling: Effect of process parameters on mechanical properties and their optimal selection. *Mater. Des.* **2017**, *124*, 143–157. [CrossRef]
6. Arora, G.; Pathak, H. Modeling of transversely isotropic properties of CNT-polymer composites using meso-scale FEM approach. *Compos. Part B Eng.* **2019**, *166*, 588–597. [CrossRef]
7. Hussin, M.S.; Hamat, S.; Ali, S.A.S.; Fozi, M.A.A.; Rahim, Y.A.; Dawi, M.S.I.M.; Darsin, M. Experimental and finite element modeling of partial infill patterns for thermoplastic polymer extrusion 3D printed material using elasto-plastic method. *AIP Conf. Proc.* **2020**, *2278*, 020011. [CrossRef]
8. Abbot, D.; Kallon, D.; Anghel, C.; Dube, P. Finite element analysis of 3D printed model via compression tests. *Procedia Manuf.* **2019**, *35*, 164–173. [CrossRef]
9. Bakradze, G.; Arājs, E.; Gaidukovs, S.; Thakur, V.K. On the heuristic procedure to determine processing parameters in additive manufacturing based on materials extrusion. *Polymers* **2020**, *12*, 3009. [CrossRef] [PubMed]
10. Pastor-Artigues, M.-M.; Roure-Fernández, F.; Ayneto-Gubert, X.; Bonada-Bo, J.; Pérez-Guindal, E.; Buj-Corral, I. Elastic asymmetry of PLA material in FDM-printed parts: Considerations concerning experimental characterisation for use in numerical simulations. *Materials* **2019**, *13*, 15. [CrossRef] [PubMed]
11. Ezeh, O.; Susmel, L. On the fatigue strength of 3D-printed polylactide (PLA). *Procedia Struct. Integr.* **2018**, *9*, 29–36. [CrossRef]
12. Song, Y.; Li, Y.; Song, W.; Yee, K.; Lee, K.-Y.; Tagarielli, V. Measurements of the mechanical response of unidirectional 3D-printed PLA. *Mater. Des.* **2017**, *123*, 154–164. [CrossRef]
13. Catana, D.; Pop, M. Studies regarding simulation process to static loading of the structures obtained from polylactic acid, 3D printed. *J. Appl. Polym. Sci.* **2021**, *138*, 50036. [CrossRef]
14. Catana, D.; Pop, M.-A.; Brus, D.-I. Comparison between the test and simulation results for PLA structures 3D printed, bending stressed. *Molecules* **2021**, *26*, 3325. [CrossRef] [PubMed]

Article

Analysis of UV Curing Strategy on Reaction Heat Control and Part Accuracy for Additive Manufacturing

Fengze Jiang *[ ] and Dietmar Drummer

Institute of Polymer Technology (LKT), Friedrich-Alexander-University Erlangen-Nuremberg,
Am Weichselgarten 10, 91058 Erlangen, Germany; dietmar.drummer@fau.de
* Correspondence: fengze.jiang@fau.de; Tel.: +49-9131-85-29736; Fax: +49-9131-85-29709

Abstract: In this research, the relationship between the curing strategies and geometrical accuracy of parts under UV light was investigated. An IR camera was utilized to monitor the process using different combinations of photosensitive resin and curing strategies. The influences of curing strategies on different material compositions were studied with single-factor analysis. With the different exposure frequencies of the UV light, the peak temperature was adjusted to avoid overheating. The three-dimensional geometry of casting tensile bars was measured to investigate the shrinkage and warpage during the curing process. Different material compositions were also selected to study the effects of the maximum temperature on the shrinkage of the parts. The findings of this work show that, with the same amount of energy input, a more fragmented exposure allows for a more controllable max temperature, while one-time exposure leads to a high temperature during the process. With the decrease of the released heat from the reaction, the shrinkage of the casting part has a slightly increasing tendency. Moreover, the warpage of the parts decreased drastically with the decrease of temperature. The addition of fillers enhances the control over temperature and increases the geometrical accuracy.

Keywords: photopolymerization; curing strategy; reaction heat; shrinkage and warpage; additive manufacturing

Citation: Jiang, F.; Drummer, D. Analysis of UV Curing Strategy on Reaction Heat Control and Part Accuracy for Additive Manufacturing. *Polymers* 2022, 14, 759. https://doi.org/10.3390/polym14040759

Academic Editors: Swee Leong Sing and Wai Yee Yeong

Received: 25 January 2022
Accepted: 13 February 2022
Published: 15 February 2022

Publisher's Note: MDPI stays neutral with regard to jurisdictional claims in published maps and institutional affiliations.

Copyright: © 2022 by the authors. Licensee MDPI, Basel, Switzerland. This article is an open access article distributed under the terms and conditions of the Creative Commons Attribution (CC BY) license (https://creativecommons.org/licenses/by/4.0/).

1. Introduction

UV curing additive manufacturing is one of the most important branches of the additive manufacturing system. With the ever increasing expansion of the UV curing system, especially for desktop stereolithography (SLA) and digital light processing (DLP) printers, UV curing-based additive manufacturing elicits much attention from both academia and industries [1]. After the development of the fast digital light synthesis (DLS) printer, UV curing printed parts have progressed even more from prototyping to being directly used industrialized parts along with those manufactured using traditional manufacturing methods [2–4]. Moreover, since the curing requires less energy and a faster reaction rate, the high efficiency and environmentally friendly characteristics are of great importance [5].

Although UV curing-based additive manufacturing has benefits such as a high resolution, smooth surface quality, and relatively fast printing, there are still several limitations that affect the printing speed [6,7]. One of the most important factors is the reaction heat. For free radical reactions, a massive amount of heat is released during curing because of the breakdown of the carbon double bond [8,9]. With the development of personalized applications, such as tooth printing, the minimization of the resin amount to avoid the production of waste and reduce the cost is greatly desirable [10]. The lesser the amount of resin, the harder it is to exchange the released heat with the surrounding liquid based on a lower heat capacity. Since the regular thermoset resin is of a low heat conductivity, the accumulated heat is problematic, including an inadequate surface quality, insufficient mechanical performance, and low geometry accuracy [11–13]. There are several solutions available on the market to avoid these issues, but they have rarely been systematically discussed. The initial temperature of resin on the mechanical properties of parts has been discussed, which could reach a better surface quality but did not clearly change the mechanical properties [8]. Researchers have tried different methods to solve this issue: the classic path is lifting the printing platform to let parts cool in the air and using a stirring bar to remix the resin tank to help heat distribution; the surface exposure method using micro shaking of the platform through a vacuum effect promotes the resin heat exchange between the printing area and non-printing area; and, from the chemical side, using a thinner layer or smaller exposure area to limit the heat release in the unit time can partially solve the problem. However, most of these solutions result in a lower printing efficiency. To maximize the printing speed with a large printing area, there is a dynamic cooling method that uses laminar circulating cooling oil which is running under the printing area to cool down the entire resin tank [6], but it requires extra structure and a circulation system. Composite resin, with different fillers inside, will also affect the absorption and reflection of the UV light; however, the majority of the research focuses more on the final part properties and the double bond conversion rate [14,15].

Shrinkage and warpage is another concern with regard to the free radical cured resin, the average shrinkage reaching between 5 and 20% depending on the selection of oligomers and monomers [16,17]. During the printing process, once a certain level of shrinkage and warpage are achieved, the distortion of the focal plane will increase the number of errors on the Z-axis that eventually deteriorate the integrity of the products. The delamination between layers generated from the shrinkage and warpage is one of the major drawbacks of UV curing additive manufacturing [18,19]. The origins of the problem can be classified into several major factors: chemical reaction, residual stress, cooling time, etc. [10,20,21]. The heat expansion and linear shrinkage after curing were discussed and a model was built [22]. The solution is to lower the intensity and slow down the average reaction rate, and also change to an optimized resin that has a lower shrinkage rate. The fillers could help decrease the shrinkage and warpage; however, information on the changing curing strategies of the filler resin is still missing [23].

In summary, the effect of the reaction heat released on the part during the process has not been fully discussed yet since the temperature increases the reactivity of the free radical reaction while at the same time dramatically decreasing the viscosity of the resin and heat expansion. The main focus of this research is to observe the in situ heat released during the printing process and compare the properties of the obtained final parts, including the part shrinkage and the warpage rate. These results could help investigate and optimize the processing parameters and promote the final properties of the part with a relatively small effect on the printing efficiency. Thus, it is essential to unearth the effect of the reaction heat during the curing process to improve the final printing accuracy.

In this paper, we first systematically analyze the effect of the curing strategies on the maximum local reaction temperature with the customized resin using an IR camera. The maximum temperature and the effects of different temperatures on the part shrinkage and warpage after the parts were cured are discussed. Moreover, two different shapes of fillers were added into the resin to investigate the basic effects of composite materials' shrinkage and warpage compared with the non-filler resin.

2. Methods

2.1. Preparation of Hybrid Resin

The matrix of the resin was prepared using aromatic urethane acrylate (3-isocyanato methyl-3,5,5-trimethylcyclohexyl isocyanate, Photomer 6628) as the oligomer, HDDA (Photomer 4017) as the reaction diluent, and Photoinitiator 1173 (Omnirad 1173). UV resin was purchased from IGM Resins, Netherlands.

Extra agent and fillers were also added to the resin based on the different demands of the prepared resins. Fumed silica (Aerosil 200, Evonik, Germany), with an average particle size of 12 nm, was used as the viscosity and thixotropic agent, while glass spheres and short glass fibers were implemented during the casting process. The glass spheres had a particle size between 30 and 50 μm, and the short glass fibers had an average length of 200 μm, as shown in Tables 1 and 2. The filler amount was set at 5% based on the testing limitation and reference [24].

Table 1. Milled glass fiber.

Filler	Glass Fiber	SEM
Brand	Taishan EMG-200	
Type of glass	E-glass	
Sizing agent	None	
Filament diameter	≤38 μm	
Average length	200 μm	
Moisture	≤0.1	
Loss on ignition	≤0.2	

Table 2. Glass sphere.

Filler	Glass Sphere	SEM	Size Distribution
Brand	Spheriglass 3000		
Type of glass	E-glass		
Particle size	30–50		
Bulk density	1.44 g/cm^3		

The resins and fillers were mixed by the centrifugal mixer (ARE-310, Thinky Inc., Laguna Hills, CA, USA) with 2000 rpm for 6 min to disperse all the components uniformly. After mixing, the mixed resin was gently filled into a syringe to prevent bubbles and was allowed to cool to room temperature. The material compositions are shown in Table 3.

Table 3. The composition of the test material (weight percentage).

	Oligomer	Monomer	Photoinitiators	Fumed Silica	Glass Sphere	Short Glass Fiber
1	58%	38%	3%	1%	-	-
2	57%	37%	3%	3%	-	-
3	56%	36%	3%	5%	-	-
4	54%	35%	3%	3%	5%	-
5	53%	34%	3%	5%	5%	-
6	53%	34%	3%	5%	-	5%

2.2. Casting Mold

The standard dog bone samples were prepared in accordance with the ISO 20753, standard type 1BB. An SLA printer was initially adopted for the male casting mold followed by the replication on PDMS to the female mold which is shown in Figure 1.

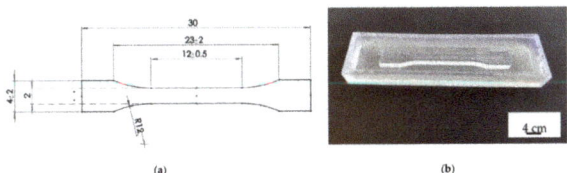

Figure 1. The casting mold of (**a**) the 1BB type geometry of tensile bar and (**b**) the male casting mold made by SLA.

2.3. Temperature Observation

The UV oven (AMP Technica AG, Germany) was equipped with three levels of light intensities. In order to assure the exposure of an equivalent amount of light energy to all samples together with the control of the reaction rate using the light intensities, the lowest level of light intensity (10 mW/cm^2) was selected. The sample was positioned in the middle of the two UV led lamps, for which the position was labeled to ensure all samples were placed in identical locations.

The IR instrument was placed at the corner of the oven with a proper angle to monitor the curing process of the tensile bar by three points (Figure 2b) which divided the bar into three parts. Moreover, the average temperature during the reaction was recorded for the calculation of heat released during the mold casting.

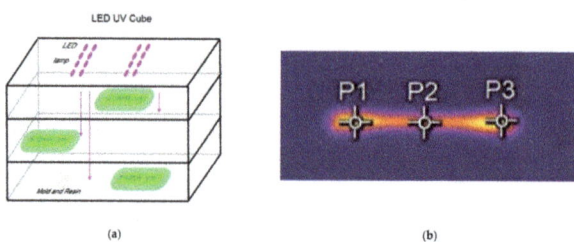

Figure 2. (**a**) Inner structure of UV oven and (**b**) sampling positions of IR camera.

2.4. Curing Strategies

During the previous research, the material had about a 3 s initiation time to generate sufficient free radicals to eliminate the oxygen and initiate the curing [25]. In comparison with the commercial printer, the ratio between the light being on and off is about 3:7; thus, we designed several groups of curing strategies to determine the effect on the average temperature of the resin in the mold. Intensive pre-testing of the suitable curing time was conducted to investigate the position where the temperature remains unchanged while oxygen has been consumed. Due to the massive heat release during curing, we decided to disjointedly expose the tensile bars to UV light with various fragmented UV exposure times, while keeping the total UV exposure time identical (8 s). The fragmented UV exposure time were 4 s, 2 s, and 1 s, which requires the repetitive exposure to UV light 2, 4, and 8 times, respectively, with 4 s time intervals, as shown in Figure 3 [25]. The detail experiment groups are shown in Table 4. The curing time was set up on the UV oven control panel and the time interval was fixed to 4 s.

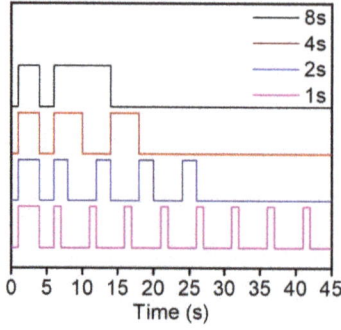

Figure 3. The curing strategies by control UV light.

Table 4. The curing strategies.

Group	Exposure Time (s)								
1	3	8	-	-	-	-	-	-	-	-
2	3	4	4	-	-	-	-	-	-	-
3	3	2	2	2	2	-	-	-	-
4	3	1	1	1	1	1	1	1	1

2.5. Shrinkage and Warpage

The pronounced shrinkage, together with the large heat release during UV curing, affects the stress distribution of the residues, which leads to the strong warpage that greatly hinders the application of UV curing resin. Thus, it is of importance to investigate the heat released from the curing, which was identified by the tensile bar geometry and warpage angle. For the tensile bar, the width, length, and thickness were measured three times for each bar, and the warpage angle was measured with the protractor. The shrinkage measurements were taken with the width and the length of the tensile bar flattened on the table as shown in Figure 4a, and the warpage was measured using the angle between the flat surface and the highest position as shown in Figure 4b.

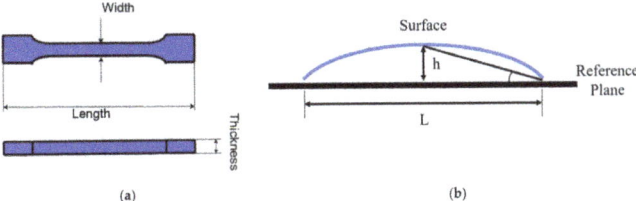

Figure 4. (a) Shrinkage measurement position and (b) warpage measurement angle.

3. Results and Discussion

3.1. Effects of Different Curing Strategies and Fumed Silica Ratio on Reaction Heat Control

Two types of resin with different amounts of fumed silica were measured, which was used to adjust the initial viscosity based on the high surface area of the nanoparticles. However, to investigate the resin at a certain viscosity, the effects of the fumed silica on temperature need to be clarified.

Although the peak temperature reveals a minute difference, the increase of temperature was significantly altered in the presence of the fumed silica, which indicates that the nanoscale silica particles may reflect the UV light to the surroundings, enhancing the curing. Comparing Figures 5a and 5b, the first peak of the temperature was reached in a shorter period of time with a higher value when 5% FS was added, except for at the 8 s exposure time, which may be due to the attainment of the highest conversion. For example, at a 4 s curing time, the temperature of 5% FS was 20 °C higher than that of 1%, while, at a 2 s curing time, the 1% FS curve revealed no peak while the appearance of a peak in 5% FS was detected.

Moreover, it was found that the curing strategies greatly affect the value of peak temperature; the was higher the frequency of the cooling intervals, the lower was the peak temperature. The above-mentioned conclusion was important for the preparation of the materials because, for low flash point materials, a lower peak temperature facilitates the reduction of heat residual stress, enabling the uniform distribution of temperature.

In general, the addition of 5% fumed silica did not drastically decrease the temperature during the curing process, and since it was utilized as the thixotropy agent for the resin, the excess addition of fumed silica decreased the flowability of the resin. However, it should be noted that the increase of the fumed silica content up to 5% results in a faster curing with a more uniform distribution [15,26].

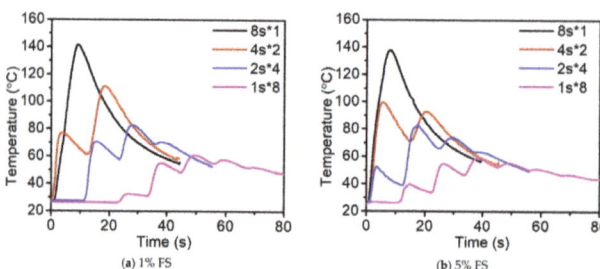

Figure 5. Effects of fumed silica amount and curing strategies on the reaction heat temperature.

3.2. Effects of Different Curing Strategies and Filler Types on Reaction Heat Control

The polarized microscope image in Figure 6 shows the distribution of fillers in the middle region of the tensile bar. Both the glass spheres and glass fibers were uniformly distributed in the bar. Moreover, in terms of the orientation of the glass fiber, as shown in Figure 6a, it was perpendicular to the paper, where the transparent holes indicate the fiber remained in place while the dark holes indicate the absence of the fiber.

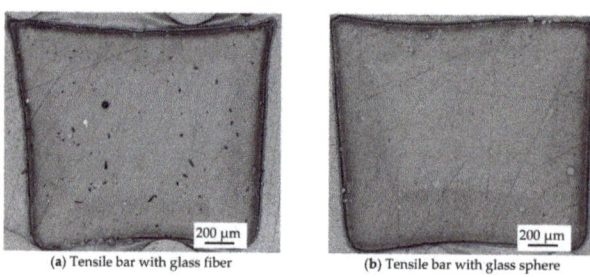

Figure 6. Polarizing microscope image of cured composite resin cross-section area.

In Figure 7, two types of fillers—glass spheres and short glass fibers—were added into the resin to study the effects of different fillers on heat control during curing. From the above-mentioned observations of fumed silica, the fillers were uniformly distributed in the resins and could maintain the dispersion for at least a month.

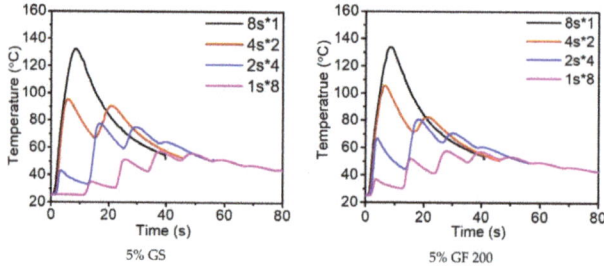

Figure 7. Effects of filler type and curing strategies on the reaction heat temperature.

In the case of an 8 s curing time, the difference between the first peak temperatures of the two fillers was insignificant; however, higher heat releases within a shorter curing time were observed for both fillers. The results may due to the fact that under a relatively long light exposure time (8 s), the impact of fillers on the curing process is negligible.

With a 4 s curing time, the glass spheres as filler demonstrated a milder heat release at the first peak (93 °C) in comparison with that of the glass fibers (105 °C). With a 2 s curing time, the glass fibers as filler revealed a higher first peak of 66 °C compared with the glass spheres' peak of 43 °C, which indicates the higher curing degree of the resin. The results may due to the inherent anisotropic structure of the glass fiber, which outcompetes the isotropic structure of the glass sphere in regard to the facilitation of curing since the oriented microstructure of the glass fiber reflects the light between fillers more efficiently.

Figure 8 presents the comparison of the peak temperatures during the curing process under different conditions. Figure 8a illustrates that with the increase of the loadings of fumed silica nanoparticles from 1% FS, the maximum reaction temperature first increased at the 3% FS loadings followed by a decrease down to its original level at the 5% FS loadings. Thus, the optimal loading of fumed silica nanoparticles that promotes the absorption and reflection of the UV light is 3% FS. Moreover, once the exposure time decreased, the decline of the average reaction temperature was significant even under identical total energy input. Regardless of the exposure time, the 3% FS loadings reveal an increase in reaction temperature and the discrepancy between 8 s and 1 s was as large as 60 °C, which decreases the thermal stress during the cooling stage and increases the printing accuracy.

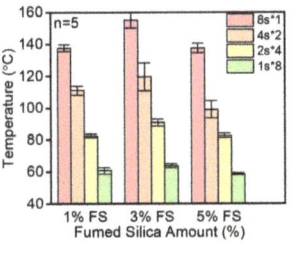
(a) Difference between different amounts of FS

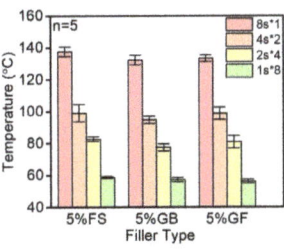
(b) Difference between different fillers

Figure 8. Comparison of the reaction temperatures using different curing strategies.

The comparison between different types of filler was also investigated. As shown in Figure 8b, the reaction heat was slightly reduced regardless of the type of filler, which further decreased the maximum reaction temperature during curing. The effects of fillers on curing depended on the shape of the fillers.; the fumed silica nanoparticles revealed the least impact while the glass spheres decreased the reaction temperature to a higher degree compared to the glass fibers. In addition to the reaction temperature, the shape of fillers may also affect the UV absorption and reflection during light exposure; however, the difference of temperature was insignificant due to the low loading of fillers (5%).

3.3. Effects of Curing Strategies on Parts Volume Shrinkage

UV-initiated radical curing is known for its shrinkage during the printing process, which drastically decreases the printing accuracy of parts and increases the printing difficulties. Thus, we tend to measure the shrinkage and warpage of the parts under different curing conditions with various curing strategies.

As shown in Figure 9, the effect of fumed silica loadings on the shrinkage of parts in three dimensions was measured. The fumed silica was initially regarded as a thixotropic agent and the suitable loading range for direct writing printing was selected from 1% to 5%. From the graph, we noticed that the increase of fume silica loadings barely impacted the length and width of the tensile bar; however, with regard to the thickness, the increase of the fumed silica content greatly increased the shrinkage. Moreover, the decrease of the fragmented exposure time from 8 s to 4 s, 2 s, and 1 s shows pronounced increases for the shrinkage of width and thickness, which may due to the fast curing of the tensile bar surface under a relatively long UV exposure time (8 s) that prevents the penetration of light into the inner layer.

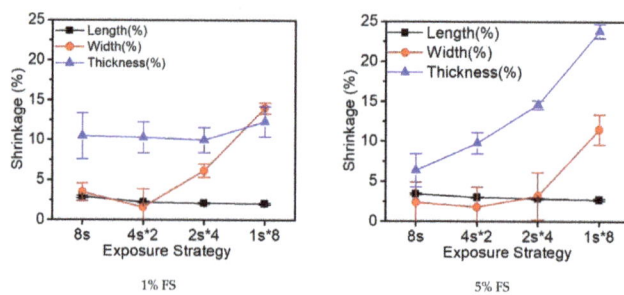

Figure 9. The shrinkage of tensile bar with different amounts of FS.

In Figure 10, the comparison between different materials and curing strategies on the shrinkage of tensile bars in three dimensions is shown. In general, the shrinkage in length was as low as 3%, which is insignificant in terms of free radical curing. Comparatively, the shrinkage in width is obvious for both the glass fibers and glass spheres and is more distinct in the latter case. In terms of thickness, due to the casting method, one side of the surface is exposed to the atmosphere, which leads to significant differences in comparison with the other two directions in shrinkage. As a result, the shrinkage in thickness increased, varying from 14% to 23% depending on the different curing strategies and materials. Conclusively, with the increase of fragmented exposure time, barely any changes were detected in length, the width increased gradually, and the discrepancy in thickness was the most significant, which may have been due to the oxygen inhibition.

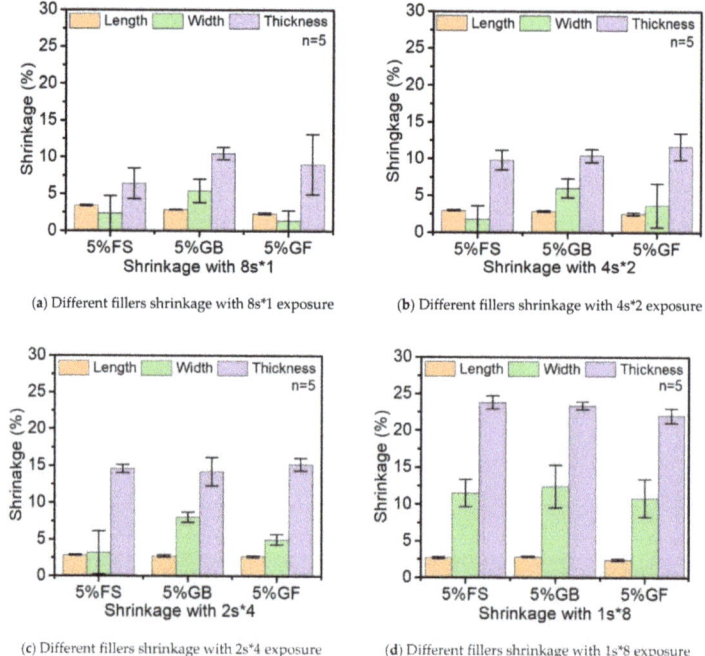

Figure 10. The shrinkage of tensile bar under four curing strategies with different fillers.

3.4. Effects of Curing Strategies on Parts Warpage

From the results of the warpage studies in Figure 11, the addition of glass spheres and glass fibers drastically decreased the warpage degree, while the glass fibers resulted in a higher degree of decreasing which may have been due to the shape and orientation of the glass fibers that enhanced the flexural strength along the axial direction. By following the direction of the mold, the glass fibers were oriented and parallel to the surface after casting.

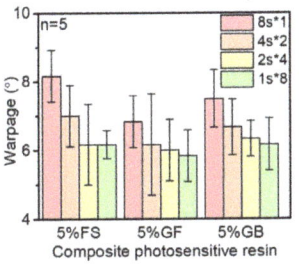 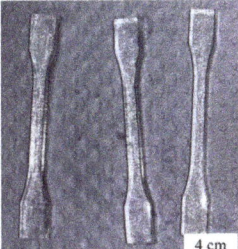

(a) Comparison of fillers under curing strategies (b) Casted tensile bars

Figure 11. The warpage of the tensile bars with different fillers.

The casting direction was extruded along with the mold, so the glass fibers were orientated by the liquid flow and kept parallel to the surface. At this position, the floating glass fibers can reinforce the tensile bar structure to resist the deformation and residual stress.

In addition, with the decrease of the fragmented exposure time, the warpage angle decreased at the same time, which means the warpage maintains the flat surface that allows a second layer of deposition during the additive manufacturing process.

4. Conclusions and Outlook

In this paper, the UV curing for casting parts was investigated to study the maximum reaction heat, shrinkage, and warpage under different curing strategies and types of fillers. The differences in curing strategies with identical total exposure times were evaluated to understand the effects on temperature control. Moreover, the effects of the type of filler were also tested to understand how the shape of fillers affects the heat release. The results show that with the increase of the loading of fumed silica, the reaction heat first increased and then decreased; however, the shrink in width and thickness was increased, especially in the fragmented exposure time of 1 s. On top of that, the addition of fillers led to a slight decrease in the maximum reaction heat, and it should be noted that the shortening of the fragmented exposure time, while keeping the total exposure time identical (8 s), accounted for the drastic decrease of the maximum reaction temperature. In light of the types of fillers, the addition of short fibers decreased the total shrinkage more than the glass spheres based on the anisotropic shape that changes the light path. Moreover, both the decreases of fragmented exposure time and the addition of fillers decreased the warpage of the parts, and the glass fibers showed a higher warpage resistance in comparison with the glass spheres.

The results of this paper provide guidance for material design and development for the UV curing resin based on different applications in terms of the model geometric accuracy. The studies of the two types of fillers on shrinkage and warpage also provide new routes for composite resins for functional printing purposes.

Author Contributions: F.J.: Conceptualization, methodology, investigations, writing—original draft, visualization. D.D.: Supervision, resources, writing—review and editing. All authors have read and agreed to the published version of the manuscript.

Funding: This research received no external funding. And the APC was funded by Friedrich-Alexander-Universität Erlangen-Nürnberg within the funding program "Open Access Publication Funding".

Institutional Review Board Statement: Not applicable.

Informed Consent Statement: Not applicable.

Data Availability Statement: Not applicable.

Acknowledgments: The authors would like to thank Zongxu Fu for the discussion and experiments. This research did not receive any specific grant from funding agencies in the public, commercial, or not-for-profit sectors. We also acknowledge financial support by Deutsche Forschungsgemeinschaft and Friedrich-Alexander-Universität Erlangen-Nürnberg within the funding program "Open Access Publication Funding".

Conflicts of Interest: The authors declare that they have no known competing financial interests or personal relationships that could have appeared to influence the work reported in this paper.

References

1. Mendes-Felipe, C.; Oliveira, J.; Etxebarria, I.; Vilas-Vilela, J.L.; Lanceros-Mendez, S. State-of-the-Art and Future Challenges of UV Curable Polymer-Based Smart Materials for Printing Technologies. *Adv. Mater. Technol.* **2019**, *4*, 1800618. [CrossRef]
2. Zhang, Y.; Dong, Z.; Li, C.; Du, H.; Fang, N.X.; Wu, L.; Song, Y. Continuous 3D printing from one single droplet. *Nat. Commun.* **2020**, *11*, 4685. [CrossRef]
3. Tumbleston, J.R.; Shirvanyants, D.; Ermoshkin, N.; Januszlewicz, R.; Johnson, A.R.; Kelly, D.; Chen, K.; Pinschmidt, R.; Rolland, J.P.; Ermoshkin, A.; et al. Continuous liquid interface production of 3D objects. *Science* **2015**, *347*, 1349–1352. [CrossRef]
4. Januszlewicz, R.; Tumbleston, J.R.; Quintanilla, A.L.; Mecham, S.J.; DeSimone, J.M. Layerless fabrication with continuous liquid interface production. *Proc. Natl. Acad. Sci. USA* **2016**, *113*, 11703–11708. [CrossRef]
5. Hopkinson, N.; Hague, R.J.M.; Dickens, P.M. (Eds.) *Rapid Manufacturing: An Industrial Revolution for the Digital Age*; John Wiley: Chichester, UK, 2006.
6. Walker, D.A.; Hedrick, J.L.; Mirkin, C.A. Rapid, large-volume, thermally controlled 3D printing using a mobile liquid interface. *Science* **2019**, *366*, 360–364. [CrossRef] [PubMed]
7. de Beer, M.P.; van der Laan, H.L.; Cole, M.A.; Whelan, R.J.; Burns, M.A.; Scott, T.F. Rapid, continuous additive manufacturing by volumetric polymerization inhibition patterning. *Sci. Adv.* **2019**, *5*, eaau8723. [CrossRef] [PubMed]
8. Steyrer, B.; Busetti, B.; Harakály, G.; Liska, R.; Stampfl, J. Hot Lithography vs. room temperature DLP 3D-printing of a dimethacrylate. *Addit. Manuf.* **2018**, *21*, 209–214. [CrossRef]
9. D'hooge, D.R.; Reyniers, M.-F.; Marin, G.B. The Crucial Role of Diffusional Limitations in Controlled Radical Polymerization. *Macromol. React. Eng.* **2013**, *7*, 362–379. [CrossRef]
10. Zhao, M.; Geng, Y.; Fan, S.; Yao, X.; Zhu, M.; Zhang, Y. 3D-printed strong hybrid materials with low shrinkage for dental restoration. *Compos. Sci. Technol.* **2021**, *213*, 108902. [CrossRef]
11. Xu, K.; Chen, Y. Photocuring Temperature Study for Curl Distortion Control in Projection-Based Stereolithography. *J. Manuf. Sci. Eng.* **2017**, *139*, 021002. [CrossRef]
12. Corcione, C.E.; Greco, A.; Maffezzoli, A. Temperature evolution during stereolithography building with a commercial epoxy resin. *Polym. Eng. Sci.* **2006**, *46*, 493–502. [CrossRef]
13. Edeleva, M.; Marien, Y.W.; Steenberge, P.H.M.V.; D'hooge, D.R. Jacket temperature regulation allowing well-defined non-adiabatic lab-scale solution free radical polymerization of acrylates. *React. Chem. Eng.* **2021**, *6*, 1053–1069. [CrossRef]
14. Atai, M.; Watts, D.C. A new kinetic model for the photopolymerization shrinkage-strain of dental composites and resin-monomers. *Dent. Mater.* **2006**, *22*, 785–791. [CrossRef]
15. Sirovica, S.; Solheim, J.H.; Skoda, M.W.A.; Hirschmugl, C.J.; Mattson, E.C.; Aboualizadeh, E.; Guo, Y.; Chen, X.; Kohler, A.; Romanyk, D.L.; et al. Origin of micro-scale heterogeneity in polymerisation of photo-activated resin composites. *Nat. Commun.* **2020**, *11*, 1849. [CrossRef]
16. Bártolo, P. (Ed.). *Stereolithography: Materials, Processes and Applications*; Springer: New York, NY, USA, 2011; ISBN 978-0-387-92903-3.
17. Fink, J.K. *Reactive polymers: Fundamentals and Applications: A Concise guide to Industrial Polymers*; William Andrew: Leoben, Austria, 2017.
18. Karalekas, D.; Aggelopoulos, A. Study of shrinkage strains in a stereolithography cured acrylic photopolymer resin. *J. Mater. Process. Technol.* **2003**, *136*, 146–150. [CrossRef]
19. Lu, B.; Xiao, P.; Sun, M.; Nie, J. Reducing volume shrinkage by low-temperature photopolymerization. *J. Appl. Polym. Sci.* **2007**, *104*, 1126–1130. [CrossRef]
20. Schmutzler, C.; Stiehl, T.H.; Zaeh, M.F. Empirical process model for shrinkage-induced warpage in 3D printing. *Rapid Prototyp. J.* **2019**, *25*, 721–727. [CrossRef]
21. Geiser, V.; Leterrier, Y.; Månson, J.-A.E. Conversion and shrinkage analysis of acrylated hyperbranched polymer nanocomposites. *J. Appl. Polym. Sci.* **2009**, *114*, 1954–1963. [CrossRef]
22. Narahara, H.; Tanaka, F.; Kishinami, T.; Igarashi, S.; Saito, K. Reaction heat effects on initial linear shrinkage and deformation in stereolithography. *Rapid Prototyp. J.* **1999**, *5*, 120–128. [CrossRef]

23. Turssi, C.P.; Ferracane, J.L.; Vogel, K. Filler features and their effects on wear and degree of conversion of particulate dental resin composites. *Biomaterials* **2005**, *26*, 4932–4937. [CrossRef]
24. Zou, H.; Wu, S.; Shen, J. Polymer/Silica Nanocomposites: Preparation, Characterization, Properties, and Applications. *Chem. Rev.* **2008**, *108*, 3893–3957. [CrossRef] [PubMed]
25. Jiang, F.; Drummer, D. Curing Kinetic Analysis of Acrylate Photopolymer for Additive Manufacturing by Photo-DSC. *Polymers* **2020**, *12*, 1080. [CrossRef] [PubMed]
26. Dubois, C.; Rajabian, M.; Rodrigue, D. Polymerization compounding of polyurethane-fumed silica composites. *Polym. Eng. Sci.* **2006**, *46*, 360–371. [CrossRef]

Article

Investigating the Potential Plasticizing Effect of Di-Carboxylic Acids for the Manufacturing of Solid Oral Forms with Copovidone and Ibuprofen by Selective Laser Sintering

Yanis Abdelhamid Gueche [1], Noelia M. Sanchez-Ballester [1,2], Bernard Bataille [1], Adrien Aubert [1], Jean-Christophe Rossi [3] and Ian Soulairol [1,2,*]

1. ICGM, University Montpellier, CNRS, ENSCM, 34000 Montpellier, France; yanis-abdelhamid.gueche@etu.umontpellier.fr (Y.A.G.); noelia.sanchez-ballester@umontpellier.fr (N.M.S.-B.); bernard.bataille@umontpellier.fr (B.B.); adrien.aubert@umontpellier.fr (A.A.)
2. Department of Pharmacy, Nîmes University Hospital, 30900 Nimes, France
3. IBMM, University Montpellier, CNRS, ENSCM, 34000 Montpellier, France; jean-christophe.rossi@umontpellier.fr
* Correspondence: ian.soulairol@umontpellier.fr

Abstract: In selective laser sintering (SLS), the heating temperature is a critical parameter for printability but can also be deleterious for the stability of active ingredients. This work aims to explore the plasticizing effect of di-carboxylic acids on reducing the optimal heating temperature (OHT) of polymer powder during SLS. First, mixtures of copovidone and di-carboxylic acids (succinic, fumaric, maleic, malic and tartaric acids) as well as formulations with two forms of ibuprofen (acid and sodium salt) were prepared to sinter solid oral forms (SOFs), and their respective OHT was determined. Plasticization was further studied by differential scanning calorimetry (DSC) and Fourier-transform infrared spectroscopy (FTIR). Following this, the printed SOFs were characterized (solid state, weight, hardness, disintegration time, drug content and release). It was found that all acids (except tartaric acid) reduced the OHT, with succinic acid being the most efficient. In the case of ibuprofen, only the acid form demonstrated a plasticizing effect. DSC and FTIR corroborated these observations showing a decrease in the glass transition temperature and the presence of interactions, respectively. Furthermore, the properties of the sintered SOFs were not affected by plasticization and the API was not degraded in all formulations. In conclusion, this study is a proof-of-concept that processability in SLS can improve with the use of di-carboxylic acids.

Keywords: selective laser sintering; di-carboxylic acids; plasticizers; solid oral forms; printability; heating temperature

1. Introduction

In the era of patient-centred medicine, the medical field is increasingly gaining interest in 3D printing [1,2]. This flexible technology enables manufacturing of complex structures that could match the anatomical and physiological needs of each patient [3]. For example, "Invisalign®" is a transparent 3D-printed orthodontic device that tailors the malocclusions of patients [4]. Recently, this innovative technology has expanded to the production of solid oral forms (SOFs) and the first Food and Drug Administration (FDA)-approved 3D printed pill, "Spritam®", was commercialized in 2015 [5]. Among the different 3D printing techniques, selective laser sintering (SLS) shows great potential to produce personalized pharmaceutical oral forms [6,7], especially porous forms such as orally disintegrating printlets (ODPs) [8,9].

SLS is a 3D printing process based on the consolidation of powder particles by selectively scanning them with a laser, such as a CO_2 laser [6]. While studies have already demonstrated the benefits of SLS to tune drug release by varying the structure of the printlets [10–12], there is still a lack of insight into the printability of pharmaceutical materials.

Thus, before exploring the endless possibilities of pharmaceutical applications with SLS, it seems necessary to master the 3D printing process. Now, efforts are being performed by pharmaceutical engineering and formulation specialists to understand the relationship between the feedstock material and the process parameters [13,14]. In SLS, prior to sintering, the powder must be heated, and this heating temperature depends on the thermal properties of the polymer (glass transition temperature and melting temperature for amorphous and crystalline polymers, respectively) [15]. Optimal heating avoids phenomena such as curling of the sintered layers due to a high thermal gradient between the high laser energy and the temperature of the powder bed [16]. Furthermore, when the powder contains more than one component, the heating temperature that ensures optimal printing can change, especially if the new component modifies the thermal properties of the mixture. For instance, it has been demonstrated that paracetamol can reduce the heating temperature of the sintering process due to its plasticizing effect [17]. Similar observations were made in hot-melt extrusion (HME) [18] and fused deposition modelling (FDM) [19], where the introduction of plasticizers or even active pharmaceutical ingredients (APIs) decreased the extrusion and printing temperatures by reducing the melt viscosity of the formulation.

In SLS, most of the energy required to consolidate the powder is provided by the heating step, and then the laser finalizes the particle coalescence to produce the printlet [20]. Plasticizers could be beneficial in lowering the heating temperature, which would protect the API from degradation. Besides thermal degradation, studies have demonstrated that recycling unsintered powder that has undergone multiple heating cycles can modify the physicochemical properties of the material, such as particle size and molecular weight [21]. Therefore, plasticizers could be used as stabilizers to conserve the initial properties of the powder by decreasing the processing temperatures.

Plasticizers are widely used in industrial manufacturing of plastics to improve the mechanical performances of the materials [22]. They are small molecules that increase the mobility of the polymer chains they are mixed with, usually by forming hydrogen bonds with the functional groups of the polymer [23]. While phthalates are one of the most frequently consumed plasticizers for the manufacturing of flexible plastics [24], they have recently been the subject of increasing concern because of their ubiquity in the environment [25] and their endocrine disrupting effects [26]. For these reasons, synthetic plasticizers are starting to be replaced by safer and more environmentally friendly plasticizers. Among these renewable and bio-sourced plasticizers are di-carboxylic acids such as succinic acid [27], fumaric acid [28] and tartaric acid [29]. Although these components are typically utilized as esters to plasticize thermoplastic polymers [30,31], they can also be used in their acid form [32].

There are several theories that explain the mechanisms behinds plasticization [33]. An example is the lubricity theory which holds that the plasticizer acts as a molecular lubricant under pressure, allowing the polymer chains to move freely [34]. In the gel theory [35], the plasticizer moves the attraction points (Van der Waals and hydrogen bonds) between the chains of the polymer which is considered as a gel, thereby increasing the mobility. The mechanistic theory is a modified version of the gel theory [36], stipulating that the plasticizer can move freely between the polymer chains by solvation / desolvation and are not fixed attraction points as suggested by the gel theory. On the other hand, the free volume theory considers that the plasticizer contributes to increase the free volume of the polymer and reduce the glass transition temperature (T_g) [37], hence providing the chains more space to move.

Although the plasticizing effect is well-known, the use of pharmaceutical excipients such as di-carboxylic acids to reduce the sintering temperature of the SLS process has not yet been explored. Thus, this work aims to investigate, as a proof of concept, the potential plasticizing effect of five structurally different di-carboxylic acids (Figure 1) by formulating them with copovidone (Kollidon VA64), one of the most investigated thermoplastic polymers in SLS for the manufacturing of solid oral forms (SOFs) [8,9,11]. As this copolymer of polyvinyl pyrrolidone and vinyl acetate only contains H-bond acceptors [38], a possible plasticizing effect can be envisaged with H-bond donors such as di-carboxylic acids. In addition, the effect of API as a plasticizer was further investigated in the case of ibuprofen. This API was chosen because its acid form possesses a carboxylic acid group that is available to act as H-bond donor as previously reported [39]. It is also a good example of a thermolabile drug with low melting and degradation temperatures which can undergo degradation when submitted to thermal stress [40]. Lastly, differential scanning calorimetry (DSC) and Fourier-transform infrared spectroscopy (FTIR) were conducted to explore the mechanisms of plasticization and the degradation of the drug was assessed by ultra-high performance liquid chromatography (UHPLC).

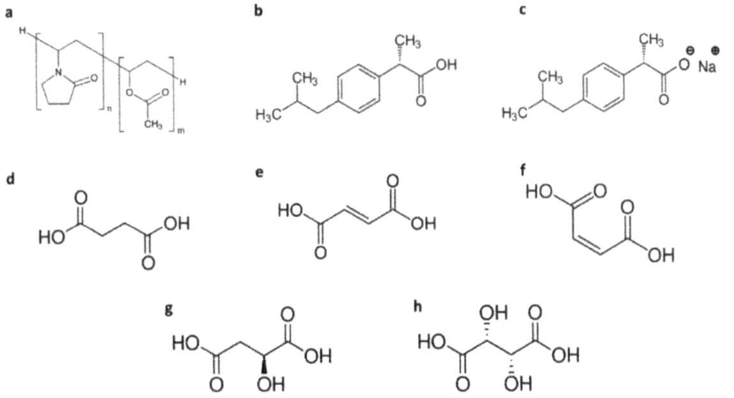

Figure 1. Structures of (a) copovidone KVA64, (b) ibuprofen acid, (c) ibuprofen sodium salt, (d) succinic acid, (e) fumaric acid, (f) maleic acid, (g) L-malic acid, (h) L-tartaric acid.

2. Materials and Methods

2.1. Materials

Kollidon® VA64 (KVA64) was donated by BASF (Ludwigshafen, Germany). Ibuprofen acid (IbuAc) and ibuprofen sodium salt anhydrous (IbuNa) were provided by Fagron (Rotterdam, Netherlands) and Sigma-Aldrich (Saint-Louis, MO, USA), respectively. Succinic acid (SA) (anhydrous, free-flowing, Redi-Dri™, ACS reagent $\geq$ 99.0%), fumaric acid (FA) ($\geq$99.0% (T)), maleic acid (MA) (ReagentPlus® $\geq$ 99% (HPLC)), L-(-)-malic acid (MLA) ($\geq$95% (titration)) and L-(+)-tartaric acid (TA) (ACS reagent $\geq$ 99.5%) were purchased from Sigma-Aldrich (Saint-Louis, MO, USA). Figure 1 illustrates the chemical structure of the different components used in the printing process.

Hydrochloric acid (37%) for the preparation of dissolution medium was purchased from Carlo Erba Reagents (Milano, Italy). Formic acid (reagent grade, $\geq$95%) and acetonitrile (HPLC gradient grade, $\geq$99%) were obtained from Sigma-Aldrich (Saint-Louis, MO, USA).

2.2. Preparation of Mixtures

Binary mixtures of KVA64/di-carboxylic acid and KVA64/drug as well ternary mixtures of copovidone/succinic acid/IbuNa were prepared. Table 1 summarizes the composition of the different mixtures. Prior to mixing, due to their large particle size, the di-carboxylic acids were ground with a pestle and mortar, then all components were sieved through a 250 μm sieve. The mixing was carried out on a 3D shaker mixer Turbula®T2F (WAB, Muttenz, Switzerland) at a speed of 49 rpm for 10 min.

Table 1. Composition of the different mixtures.

Mixtures	KVA64	IbuAc	IbuNa	SA	FA	MA	MLA	TA
KVA64/IbuAc	95%	5%						
KVA64/IbuNa	95%		5%					
KVA64/SA	95%			5%				
KVA64/FA	95%				5%			
KVA64/MA	95%					5%		
KVA64/MLA	95%						5%	
KVA64/TA	95%							5%
KVA64/SA10	90%			10%				
KVA64/SA15	85%			15%				
KVA64/SA20	80%			20%				
KVA64/IbuNa/SA	90%		5%	5%				
KVA64/IbuNa/SA10	85%		5%	10%				
KVA64/IbuNa/SA15	80%		5%	15%				
KVA64/IbuNa/SA20	75%		5%	20%				

KVA64: Kollidon VA64, IbuAc: ibuprofen acid, IbuNa: ibuprofen sodium salt, SA: succinic acid, FA: fumaric acid, MA: maleic acid, MLA: malic acid, TA: tartaric acid.

2.3. Printing of Solid Oral Forms

First, OnShape®(Onshape, Boston, MA, USA), an online computer-aided design (CAD) software and Slic3r® 1.2.9, an open-source software, were used for the design and slicing of the cylindrical SOF (10 mm diameter × 3 mm height), respectively. Then, 300 g of each powder was loaded into the Sharebot® SnowWhite 3D SLS printer (Sharebot, Nibionno, Italy) and thirty-six SOFs were launched for batch printing. Before printing, the powder was heated during 30 min and the temperature mode was set to *powder temperature*. While the laser power, scan speed and layer thickness were set constant (25%, 35,000 pps and 100 µm, respectively), the powder temperature was modified as this is a key parameter for this study. It was first set to a low value empirically based on the glass transition temperature (T_g) of the powder mixture and then incremented by 5 °C until the optimal heating temperature was attained (Table 2). The optimal heating temperature (OHT) corresponded to the minimum temperature at which all SOFs were completely printed without curling of the sintered layers in a reproducible manner (three times).

Table 2. Optimal heating temperatures for the different powders.

Powders	Optimal Heating Temperature (°C)
KVA64	110
KVA64/IbuAc	70
KVA64/IbuNa	110
KVA64/SA	95
KVA64/FA	105
KVA64/MA	105
KVA64/MLA	105
KVA64/TA	110
KVA64/SA10	85
KVA64/SA15	80
KVA64/SA20	80
KVA64/IbuNa/SA	95
KVA64/IbuNa/SA10	85
KVA64/IbuNa/SA15	80
KVA64/IbuNa/SA20	80

After sintering with a CO_2 laser (λ = 10.6 µm), SOFs were removed from the printing bed and brushed to remove their powder excess.

2.4. Differential Scanning Calorimetry (DSC)

DSC was used to determine the melting point (or glass transition temperature) of the individual components (copovidone, drugs and di-carboxylic acids) and the different physical mixtures prepared. Samples of 5–10 mg were placed in sealed aluminum pans and heated from 25 °C to 200 °C at 10 °C/min with a DSC 4000 (Perkin Elmer, Waltham, MA, USA). A heat-cool-heat cycle method was used to remove the thermal history of copovidone. Nitrogen was employed as the purge gas with a flow rate of 20 mL/min. Data collection and analysis were carried out with Pyris Manager software (Perkin Elmer, Waltham, MA, USA). The glass transition temperature measurements were realized in triplicate and the results were expressed as the mean value ± standard deviation.

2.5. Thermogravimetric Analysis (TGA)

TGA was used to characterize the degradation profile and determine the 2% degradation point ($Td_{2\%}$) of the different components (copovidone, drugs and di-carboxylic acids). The measurements were performed with a TGA Q50 (TA instruments, Waters Corporation, New Castle, DE, USA) from 30 °C to 700 °C at a heating rate of 15 °C/min under air flow. Platinum pans were used with an average sample weight of 10 mg. Data analysis was conducted using Universal Analysis 2000 (TA instruments, Waters Corporation, New Castle, DE, USA).

2.6. Fourier-Transform Infrared Spectroscopy (FTIR)

Infrared spectrophotometer Vector 22 FTIR (Bruker, Billerica, MA, USA) was employed to investigate potential hydrogen bond interactions formation during SLS. The absorbance of individual components (copovidone, drugs and di-carboxylic acids) and sintered SOFs was recorded from 4000 to 400 cm^{-1} at room temperature and averaged over 32 scans at 2 cm^{-1} resolution. Disks of 100 mg were prepared by mixing and then compressing 10 mg of the sample (or 0.5 mg for the drugs and the di-carboxylic acids) with Q.S. (Quantum satis) of anhydrous potassium bromide (previously dried in the oven at 100 °C for 30 min). The FTIR spectrums were treated using OPUS 6.5 infrared software (Bruker, Billerica, MA, USA).

2.7. X-ray Powder Diffraction (XRPD)

The solid state of the materials used in this study and the sintered SOFs was characterized using a Bruker D8 Advance diffractometer (Bruker, Billerica, MA, USA) and monochromatic Cu Kα1 radiation ($\lambda\alpha$ = 1.5406 Å, 40 kV and 40 mA). The angular range of data recorded was 2–70° 2θ, with a stepwise size of 0.02° and a speed of 0.1 s counting time per step, using LINXEYE detector 1D.

2.8. Weight, Hardness and Disintegration Time of the Sintered SOFs

For each formulation, the weight and hardness of ten SOFs were determined using an Adventurer® precision electronic balance (OHAUS, Parsippany, NJ, USA) and a Sotax Multitest 50FT (Sotax AG, Switzerland), respectively.

Disintegration tests were performed on a Sotax DT50 disintegration apparatus (Sotax AG, Switzerland) with distilled water (800 mL) at 37 °C following the European Pharmacopeia guidelines [41]. For each powder, six SOFs were tested simultaneously. The disintegration time was reached when no residue was present at the bottom of the test basket. The results were reported as a mean value ± standard deviation.

2.9. Drug Release of the Sintered SOFs

A dissolution test was carried out for SOFs containing IbuNa or IbuAc with a Pharma Test DT70 dissolution tester (Hainburg, Germany) using a paddle-type apparatus (European Pharmacopeia) [42]. For each formulation, three SOFs were individually placed in the dissolution vessels each containing 800 mL of 0.1 M HCl and stirred at 100 rpm and 37 ± 0.5 °C. Samples were automatically analyzed every 5 min using a continuous flow system connected to an 8 cell Specord 250 UV/Vis spectrophotometer (Analytik Jena, Germany) at a wavelength of 268 nm. The results were expressed as mean values ± standard deviation.

2.10. Drug Content of the Sintered SOFs

For each formulation, three SOFs were dissolved in 100 mL of a mixture (40%/60%) of solvent A (distilled water/formic acid (1%, v/v)) and solvent B (acetonitrile). Samples of the solutions were then diluted and the concentration of the drug was determined by ultra-high performance liquid chromatography (UHPLC) using a UHPLC-DAD system. This consisted of a Thermo Scientific™ Dionex™ UltiMate™ 3000 BioRS equipped with a WPS-3000TBRS autosampler and a TCC-3000RS column compartment set at 35 °C (Thermofisher Scientific, Waltham MA, USA). The system was operated using Chromeleon 7 software. An Accucore C18 column (2.6 μm, 100 mm × 2.1 mm) combined with a security guard ultra-cartridge (Phenomenex Inc., Torrance CA, USA) was used. An isocratic binary solvent system was utilized, consisting of solvent A and solvent B (40%A, 60%B). The flow rate of the mobile phase was 1.5 mL/minute, and the injection volume was 50 μL. Quantitative analysis of IbuAc and IbuNa in the SOFs was carried out using an external standard method. The calibration curve for each form of the drug was constructed using 5 different standard levels of the corresponding form in the concentration range 1–20 mg/L. The peak of ibuprofen was monitored at 258 nm.

2.11. Statistical Analysis

The effect of formulation composition on the glass transition temperature, weight, hardness and disintegration time of SOFs was analyzed statistically. One-way analysis of variance (ANOVA) in conjunction with Tukey's HSD (honestly significant differences) test were used to determine the statistical significance of the differences among the groups ($p < 0.05$).

3. Results and Discussion

For this study, different di-carboxylic acids were investigated as potential plasticizers. In order to assess how structural changes such as the presence of a double bond, orientation of the carboxylic acid groups in the molecule or the presence of hydroxyl groups can affect the plasticizing effect, succinic acid (butanedioic acid) was compared with fumaric ((2E)-but-2-enedioic), maleic ((2Z)-but-2-enedioic), malic (2-hydroxybutanedioic) and tartaric (2,3-dihydroxybutanedioic) acids. In addition, IbuAc and IbuNa were used to explore the potential effect on plasticization of switching an API from its acid form to its salt form.

3.1. Thermal Analysis

One of the most common ways to demonstrate a plasticizing effect is to analyse the samples by DSC and observe whether a decrease in the glass transition temperature (T_g) of the polymer is detected [43]. In this study, T_g of the different mixtures were measured from the second heating scan after removal of their thermal history [17]. After the first heating scan, the components present in all formulations were dissolved in the polymeric matrix and amorphized, hence no endothermic melting peaks were observed. Therefore, DSC thermograms of the physical mixtures containing copovidone and drugs or acids revealed only the T_g. Table 3 shows the glass transition temperatures of copovidone and the different prepared mixtures.

Table 3. Thermal properties of the different components and mixtures.

Powders	T_g (°C)	T_m (°C)	$Td_{2\%}$ (°C)
KVA64	99.30 ± 1.27	/	302.54
IbuAc	/	79.51	137.12
IbuNa	/	164.61	234.09
SA	/	190.42	166.86
FA	/	299.01	193.77
MA	/	146.65	142.15
MLA	/	106.67	165.79
TA	/	174.99	190.74
KVA64/IbuAc	80.69 ± 0.47	/	235.97
KVA64/IbuNa	93.15 ± 1.41	/	278.71
KVA64/SA	84.59 ± 0.86	/	237.92
KVA64/FA	84.25 ± 1.54	/	256.67
KVA64/MA	85.68 ± 0.24	/	207.91
KVA64/MLA	84.26 ± 1.35	/	239.32
KVA64/TA	89.55 ± 1.27	/	230.65
KVA64/SA10	74.40 ± 1.18	/	218.57
KVA64/SA15	66.84 ± 1.88	/	208.37
KVA64/SA20	57.51 ± 0.89	/	195.95
KVA64/IbuNa/SA	75.39 ± 1.29	/	237.82
KVA64/IbuNa/SA10	63.90 ± 0.34	/	217.75
KVA64/IbuNa/SA15	58.56 ± 1.77	/	197.90
KVA64/IbuNa/SA20	45.07 ± 1.05	/	186.50

T_g: glass transition temperature, T_m: melting temperature, $Td_{2\%}$: temperature at 2% of degradation.

Results in Table 3 show a decrease in the glass transition temperature of copovidone from 99.3 °C to 93.2 °C and 80.7 °C in the presence of IbuNa and IbuAc, respectively. The greater decrease in T_g observed with IbuAc may be due to the formation of hydrogen bonds between the copovidone and the carboxylic acid group of IbuAc which induces an increase in the mobility of the system [44]. Table 3 shows that adding the different di-carboxylic acids at rate of 5% to copovidone provokes a reduction in T_g. On one hand, T_g decreased to 89.6 °C after TA was incorporated. On the other hand, mixtures with the other acids (SA, FA, MA and MLA) presented a lower and similar T_g (84.3–85.7 °C) with no statistically significant difference. In addition, as the proportion of succinic acid increased, the T_g reduced proportionally to 57.5 °C for 20% of SA. Furthermore, ternary mixtures containing copovidone, SA and IbuNa showed a lower T_g compared to their binary counterpart mixtures with the same percentage of SA, which could imply a synergistic effect of SA and IbuNa on plasticization.

The advantage of using plasticizers is their ability to reduce the optimal heating temperatures (OHT). This effect can be evidenced by a solidification of the powder bed when processing the mixture at the same OHT for the polymer. This implies that the heating temperature is too high for the mixture as it exhibits a high flow and consolidates in a block before the laser even starts sintering. The heating should only approach the powder temperature to the point when the material starts to flow without inducing fluidization [45]. In contrast, when a powder containing a non-plasticizing component is processed at a temperature inferior to the OHT of the raw polymer, a curling typically occurs.

Based on these observations, the OHT for the different powders were identified as the minimum heating temperature at which no curling or powder solidification were observed (Table 2). It can be pointed out that despite a decrease in the T_g when both IbuNa and TA were mixed with KVA64, the OHT of the mixture powders was not different from the OHT for pure KVA64. For the other acids (SA, FA, MA, MLA), as well as IbuAc, a decrease in OHT was observed (Table 2). While the addition of 5% of IbuAc favoured a decrease in OHT by 40 °C, a reduction of 15 °C was observed when 5% of SA was added. A decrease of only 5°C was observed after the addition of 5% of FA, MA or MLA.

These differences demonstrate that di-carboxylic acids have a different effect on processability even though each of SA, FA, MA and MLA show a similar decrease in T_g during DSC analysis. Furthermore, as the percentage of SA increased, the OHT decreased further until it reached a minimum of 80 °C for 15% of SA (Table 2). Above this ratio, the OHT was not affected even though T_g continued to decrease at 20% of SA, demonstrating a saturation of plasticization at 15% of SA. This saturation effect has been previously reported with other plasticizers. For example, Aydin et al. demonstrated that increasing the mannitol ratio above 5% does not improve plasticization of starch films [46]. Moreover, IbuNa did not affect the processability when it was introduced in the mixtures of KVA64 and SA, as the OHT were the same for binary and ternary mixtures at an equivalent ratio of succinic acid (Table 2). Hence, no synergistic effect between IbuNa and succinic acid on processability was demonstrated.

Among the different excipients used, SA demonstrated the highest plasticizing efficiency. Therefore, the linearity of the T_g and OHT of the binary and ternary mixtures at four different succinic acid ratios (0, 5, 10 and 15%) was evaluated in the same manner as in previous work with binary mixtures of KVA64 and paracetamol at different ratios [17]. Figure 2 confirms the existence of a linear relationship with high correlation coefficients (R^2) 0.9945 and 0.9997 for both the binary and ternary mixtures, respectively. It is important to note that this linearity is not maintained above 15% of succinic acid. As the slope and the ordinate at the origin are not the same for both systems, it could be postulated that the linear regression equation relating T_g and OHT is only constant within a mixture (binary or ternary) of the same composition at different ratios of one component.

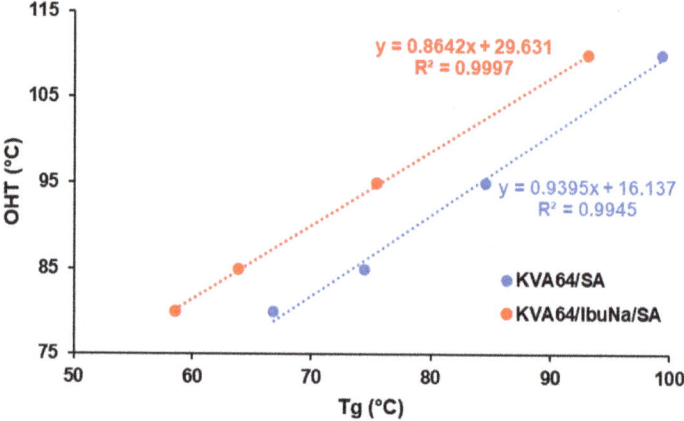

Figure 2. Relationship between glass transition temperatures and optimal heating temperatures at different ratios of succinic acid for mixtures of KVA64/SA and mixtures of KVA64/IbuNa/SA.

The method applied to measure the glass transition temperatures shows limits, as it requires, firstly, to be heated to 200 °C to remove the thermal history of KVA64, which exhibits an important endothermic peak at ~170–200 °C despite its amorphous state (Supplementary material: Figure S1). In contrast, the heating temperatures used during the SLS process did not exceed 110 °C (Table 2). Therefore, the plasticization that occurs during the first heating of DSC does not perfectly reproduce the interactions that take place during the SLS heating step. The higher temperatures used in DSC favour more contact between the small molecules and the polymer, and the generation of chemical interactions [47]. Hence, the measured glass transition temperatures may be overestimated. This could explain the reduction of T_g even with components that do not exhibit a plasticizing effect in SLS. However, these results are still of interest and serve as initial comparison of the effect of different plasticizers and ratios on the SLS printing temperature.

Table 3 shows that the melting of the different di-carboxylic acids and both forms of ibuprofen takes place at temperatures higher than the OHT of the respective mixtures. This confirms that powder consolidation during the heating step was not influenced by melting of the added excipients or drugs, but only dependant on the glass transition of the polymer as it accounts for most of the powder bed (75–95%). Furthermore, the solidification of an amorphous polymer depends on its glass transition temperature [15,16,48].

Temperatures for which 2% of the component was degraded were evaluated with TGA (Table 3). Copovidone degraded by 2% at 302 °C whereas $Td_{2\%}$ of the di-carboxylic acids were between 142 and 194 °C. Furthermore, IbuNa exhibited a higher thermal stability ($Td_{2\%}$ = 234 °C) compared to IbuAc ($Td_{2\%}$ = 137 °C). The $Td_{2\%}$ of the prepared mixtures (Table 3) were found to be inferior compared to pure copovidone due to the addition of more thermolabile drugs and di-carboxylic acids. The heating temperatures in SLS were largely below the degradation points of the different components, but the CO_2 laser can also degrade them, particularly IbuAc. Thus, UHPLC assays were conducted to further investigate this.

TGA analysis showed that the water content (WC%) of the different acids and IbuAc was inferior to 0.5%. Meanwhile, KVA64 and IbuNa showed high WC%: 2.35% and 14.66%, respectively. The anhydrous form of racemic (R,S)-(±)-ibuprofen sodium salt is highly unstable and converts rapidly to the dihydrate form when exposed to the humid environment [49], which explains the high water content corresponding to the water of crystallization. A dehydration endothermic peak was also observed in the DSC thermogram of IbuNa at 102.6 °C (Supplementary material: Figure S2). Despite the well-known plasticizing properties of water [50,51], it did not play a role in lowering the optimal heating temperature when it was bound to sodium salt ibuprofen.

3.2. FTIR Analysis

FTIR was used to explore the potential formation of H-bonds between the different materials used in the formulations studied and to provide insight into the plasticization mechanisms that occur during the heating step in SLS. These bonds could form between the C=O groups of the copovidone and the OH groups of carboxylic acids, as previously demonstrated by Hurley et al. [52]. KVA64 has two strong main peaks at 1740 cm^{-1} (C=O stretch of the vinyl acetate) and 1683 cm^{-1} (C=O stretch of the tertiary amide) as shown in Figure 3a. The variation of these two peaks was studied as copovidone is the major component in the formulations used in this work. FTIR spectrum of APIs (Figure 3a) showed a peak at 1549 cm^{-1} for IbuNa due to the C=O stretch of the carboxylate group and a sharp peak at 1721 cm^{-1} for IbuAc corresponding to the C=O stretch of the carboxylic acid group. Both C=O peaks of KVA64 did not shift to lower or higher frequencies when the polymer was sintered with both drugs, but they broadened and decreased in intensity. The broadening was more pronounced with the tertiary amide C=O peak (1683 cm^{-1}) and when IbuNa was introduced. This indicates the presence of interactions between the polymer and both drugs.

Fumaric acid and maleic acid were characterized by their C=O carboxylic stretch peaks at 1674 and 1706 cm^{-1}, respectively (Figure 4a). Malic acid presents a strong peak at 1731 cm^{-1} due to its COOH group and a large band from 2700 to 3700 cm^{-1} due to its hydroxyl group (Figure 4a). Furthermore, tartaric acid shows a small C=O stretch peak at 1738 cm^{-1} and a large band from 2700 to 3700 cm^{-1} due to its two hydroxyl groups (Figure 4a). In the spectrums of SOFs, the peak of KVA64 at 1683 cm^{-1} broadened (Figure 4b), which could be due to the formation of H-bonds. These variations could also be explained by the overlapping of the C=O peaks of di-carboxylic acids with the characteristic peaks of KVA64.

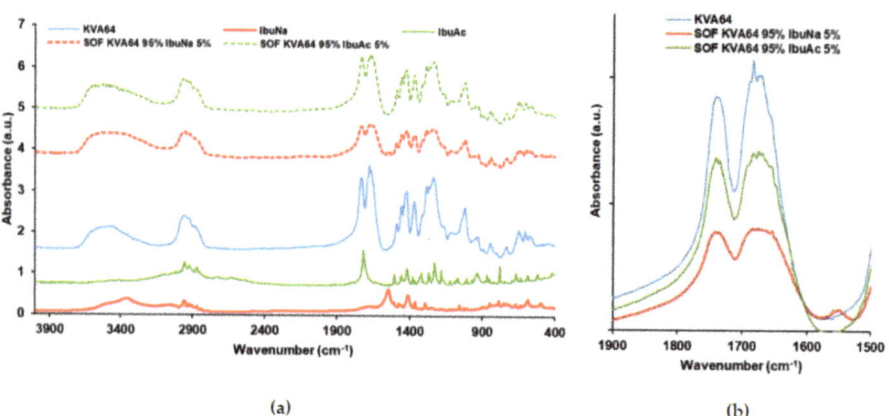

Figure 3. FTIR spectrums of KVA64, IbuNa, IbuAc and printed SOFs: (**a**) from 400 to 4000 cm^{-1}, (**b**) from 1500 to 1900 cm^{-1}.

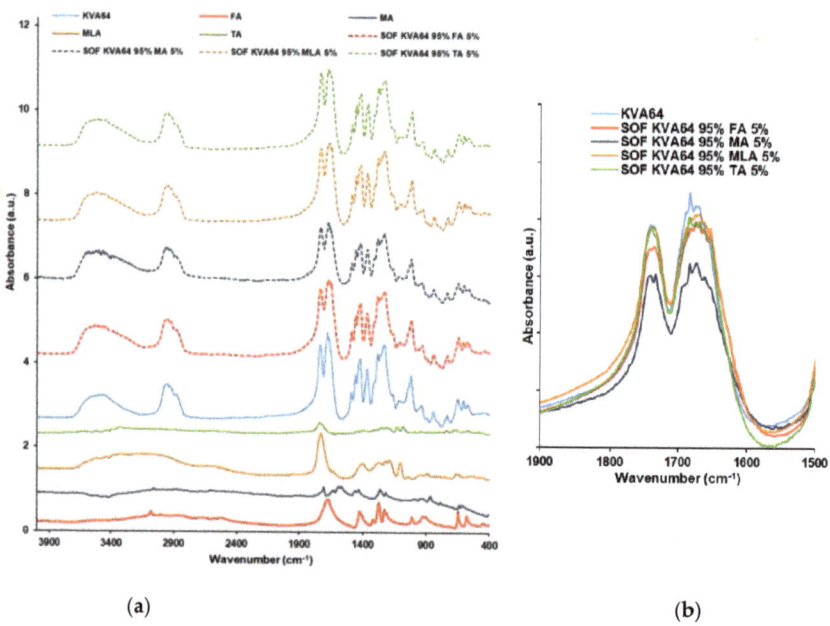

Figure 4. FTIR spectrums of KVA64, FA, MA, MLA, TA and printed SOFs: (**a**) from 400 to 4000 cm^{-1}, (**b**) from 1500 to 1900 cm^{-1}.

Succinic acid exhibited in FTIR a characteristic peak at 1681 cm^{-1} due to C=O stretch of its carboxylic group, which overlaps with the characteristic peaks of KVA64 (Figure 5a). FTIR spectra of SOFs prepared with different ratios of SA showed a broadening of the C=O amide peak of KVA64. This could be due to H-bond formation or overlapping of the bands of the two components.

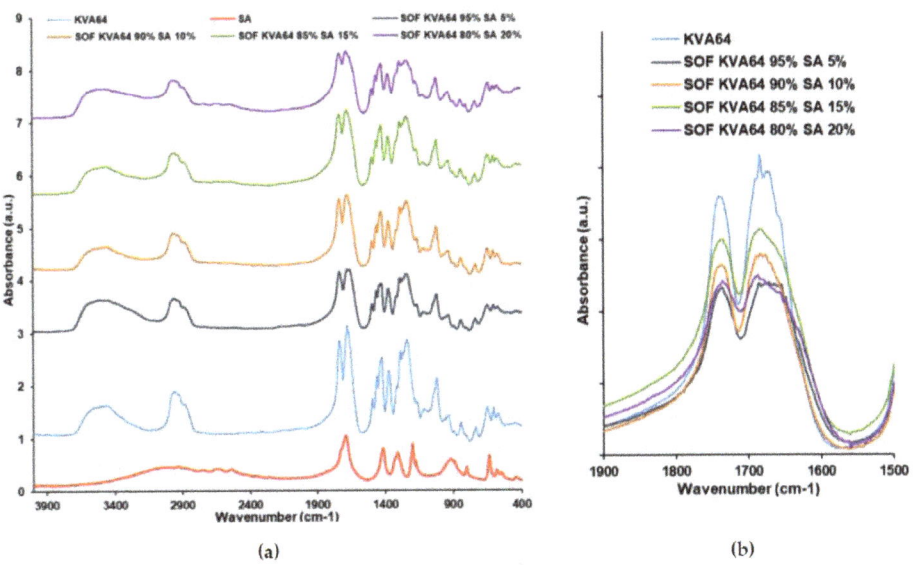

Figure 5. FTIR spectrums of KVA64, SA and printed SOFs: (**a**) from 400 to 4000 cm^{-1}, (**b**) from 1500 to 1900 cm^{-1}.

The FTIR spectra of the SOFs prepared with KVA64, IbuNa and SA also showed a broadening of the C=O amide peak of copovidone (Figure 6b). Notably, although the C=O stretch peak of IbuNa at 1549 cm^{-1} was found in all the SOFs, its intensity decreased with the proportion of succinic acid (Figure 6a), suggesting that a high content of SA promotes the amorphization of IbuNa.

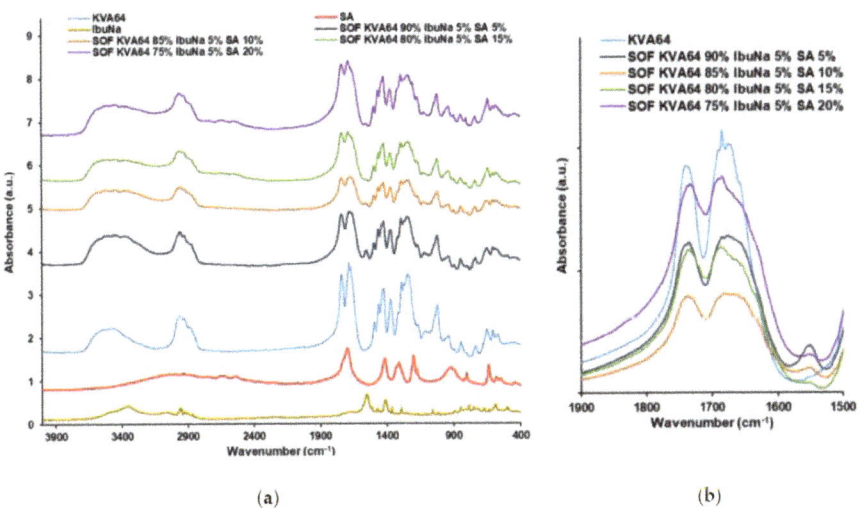

Figure 6. FTIR spectrums of KVA64, SA, IbuNa and printed SOFs: (**a**) from 400 to 4000 cm^{-1}, (**b**) from 1500 to 1900 cm^{-1}.

Overall, all the SOFs prepared with mixtures of KVA64, di-carboxylic acids and/or drugs display interactions especially at the wavenumber 1683 cm^{-1}, which corresponds to the C=O amide peak of copovidone. This group is more reactive and more susceptible to form H-bonds than the C=O vinyl acetate function, as previously reported by Yuan et al. [53]. Although no change in frequency has been observed, variations in the width and intensity could also be interpreted as the formation of intermolecular hydrogen bonds [46]. The decrease in the intensity and the width of the C=O amide peak may indicate that the interactions between the copovidone molecules are being replaced by interactions between the polymer and the other SOFs components. Nevertheless, the FTIR data did not allow differentiation between the distinct components in terms of plasticization. For instance, all of the di-carboxylic acids and drugs exhibited interactions, despite having a different effect on the optimal heating temperature. Moreover, as the proportion of succinic acid increased, there was no proportional broadening of the peak (Figures 5b and 6b).

For the IbuNa, even with an absence of H-bond donors in its structure, interactions were found with the C=O group of KVA64. This may be due to the important moisture content present in the ibuprofen sodium salt that could participate in H-bond formation [54].

As noted by Matet et al. [55], FTIR is not a discriminatory technique and could not reveal differences between the different polymer/plasticizers mixtures. Moreover, the high hygroscopicity of copovidone [56] and the overlapping of the peaks are limits for the interpretation of FTIR results. H-bonds are temporary bonds that could evolve as function of the applied temperatures [57]. Hence, further work could involve conducting FTIR at the temperatures of the heating process [58] to mimic the interactions between the components under real sintering conditions.

3.3. Solid State Analysis

XRPD analysis was carried out to study the solid state of the different components and the potential solid-state transitions due to the sintering process. All the used materials are crystalline except copovidone which is amorphous and did not expose any crystalline peaks on XRPD. X-ray diffractograms of drugs (Figure 7a) display the characteristic crystalline peaks of the racemic form for both ibuprofen acid and ibuprofen sodium salt dihydrate, in agreement with the literature data [49,59]. However, these crystalline peaks were absent in the diffractograms of the sintered SOFs (Figure 7a), indicating an amorphization of the drugs. Previous studies have already demonstrated that SLS could produce amorphous solid dispersions, due its two thermal steps (heating and laser scanning) [60,61].

Figure 7b exhibits the effect of sintering on the initially crystalline di-carboxylic acids. While MA and MLA underwent amorphization as demonstrated by the absence of peaks in the X-ray diffractograms of the SOFs prepared with these acids. For the case of SOFs prepared with FA and TA, some peaks were still distinguishable but with reduced intensity, suggesting a partial amorphization of the fumaric and tartaric acids. The differences observed in the rate of amorphization could be correlated with the melting points of the di-carboxylic acids. As FA and TA present the higher melting points, higher temperatures will be needed to dissolve completely in the polymeric matrix.

X-ray diffractogram of the succinic acid (Figure 7c) shows characteristic peaks of the form β [62]. As previously seen for MA and MLA, no peaks were observed in the diffractogram of the SOF sintered with 5% of SA, implying an amorphization of the acid. However, the SOFs prepared with higher ratio of SA exhibited peaks at 20.1, 22.1, 26.2, 27.3, 31.6, 32.6° and their intensity increased proportionally with the percentage of succinic acid incorporated. It is interesting to note that two peaks (22.1 and 27.3°) did not appear in the diffractogram of pure SA. These two peaks are specific to the form α, which is favoured by high temperatures, as suggested by Yu et al. [62]. This implies that in addition to a partial amorphization of the succinic acid, sintering promoted the recrystallisation into the form α.

Regarding the SOFs prepared with the ternary mixture of copovidone, IbuNa and SA (Figure 7d), partial amorphization of the β-succinic acid and recrystallisation into the form α was also observed at the SA ratios of 10, 15 and 20%. As for the IbuNa, three small peaks characteristic of the API (17.5, 18.3, 19.0°) and two other peaks that were not in the diffractogram of the pure drug (22.3 and 22.6°) were present in the diffractogram of SOFs sintered with 5% of SA. This could be explained by the dehydration of the ibuprofen sodium salt dihydrate during the SLS process which could induce the rearrangement of the crystal structure as previously demonstrated by Censi et al. [49]. However, only the peak at 18.3° remained at high ratios of SA (15 and 20%). These results are in agreement with the FTIR observations, as the amorphization is stimulated at higher percentage of succinic acid.

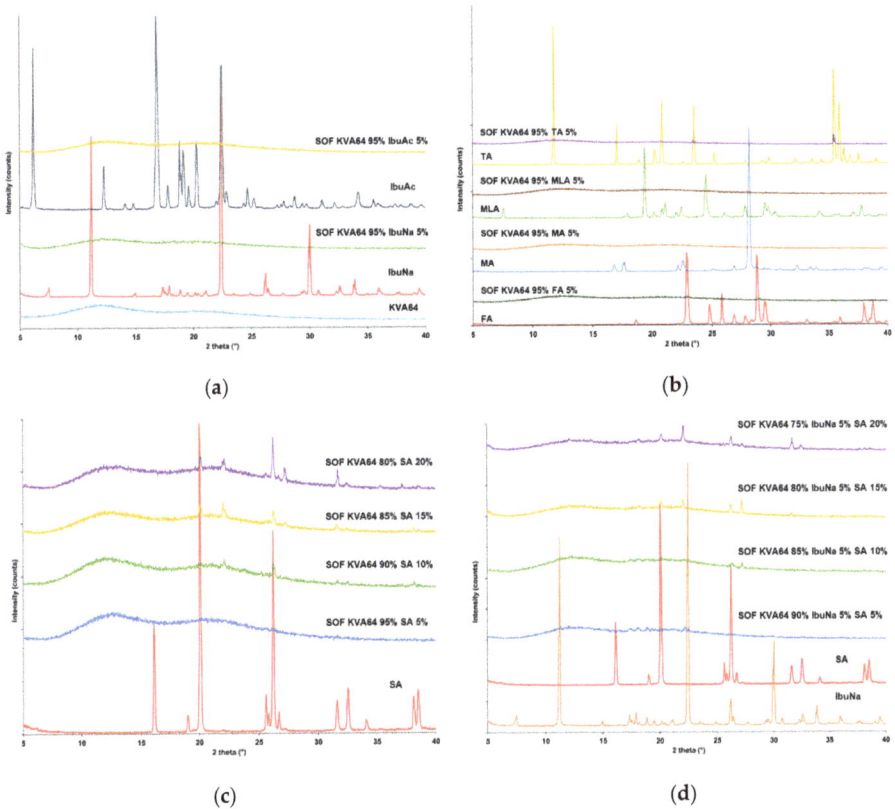

Figure 7. XRPD patterns: (**a**) KVA64, IbuNa, IbuAc and their sintered SOFs, (**b**) FA, MA, MLA, TA and their sintered SOFs, (**c**) SA and its sintered SOFs, (**d**) IbuNa, SA and their sintered SOFs.

3.4. Characterisation of the Sintered SOFs

The weight of the SOFs printed with the different powders varied between 129 and 170 mg (Table 4). The average weight of the SOFs decreased significantly with the introduction of IbuAc and increased significantly when IbuNa was incorporated. Among the different di-carboxylic acids, only SA and MA reduced the weight of SOFs significantly when they were incorporated. SOFs prepared with formulations of copovidone, SA and IbuNa exhibited a significant lower weight compared to the corresponding binary mixtures. However, increasing the succinic acid ratio did not show a significant influence on the weight variation. Table S1 (Supplementary material) shows that these variations in weight were not solely influenced by the powder compactness. This was confirmed by the observation of the SOF's vertical sections by scanning electron microscopy (SEM) (Supplementary material: Figure S3). SOF printed with the mixture of KVA64 and the plasticizing SA showed a similar structure to the SOF produced with KVA64 and the non-plasticizing TA. Further work is ongoing to help understand better these observations.

Table 4. Weight, hardness and disintegration time of the sintered SOFs.

Powders	Weight (mg)	Hardness (N)	Disintegration Time (s)
KVA64	153.0 ± 2.5	38.6 ± 4.5	73 ± 14
KVA64/IbuAc	132.2 ± 4.6	15.8 ± 3.8	24 ± 7
KVA64/IbuNa	170.8 ± 7.6	29.8 ± 2.5	58 ± 8
KVA64/SA	140.1 ± 2.1	25.6 ± 1.5	39 ± 4
KVA64/FA	149.1 ± 4.8	33.0 ± 1.7	43 ± 8
KVA64/MA	142.2 ± 1.8	24.2 ± 4.0	30 ± 8
KVA64/MLA	147.4 ± 1.1	31.9 ± 1.7	57 ± 9
KVA64/TA	166.0 ± 7.6	33.8 ± 4.2	43 ± 5
KVA64/SA10	138.4 ± 1.8	24.1 ± 2.3	25 ± 3
KVA64/SA15	139.2 ± 2.3	23.5 ± 3.0	21 ± 4
KVA64/SA20	138.9 ± 1.6	25.2 ± 2.4	23 ± 3
KVA64/IbuNa/SA	133.9 ± 1.7	16.6 ± 1.8	24 ± 3
KVA64/IbuNa/SA10	129.9 ± 1.4	16.5 ± 1.2	25 ± 3
KVA64/IbuNa/SA15	131.3 ± 2.5	18.8 ± 1.8	34 ± 5
KVA64/IbuNa/SA20	128.9 ± 1.7	18.0 ± 1.0	30 ± 6

Table 4 shows that all printed SOFs presented a low hardness (<40 N), which was expected because they were printed at a high scanning speed (35,000 pps) [8]. Forms printed only with KVA64 exhibited the higher hardness values (38.6 N). The mechanical properties of SOFs decreased significantly after the introduction of drugs, most importantly with IbuAc. As for the di-carboxylic acids, only SA, MA and MLA showed a significant negative effect on hardness. The hardness was not significantly affected by increasing the ratio of succinic acid, whether in the binary mixtures (KVA64/SA) or the ternary mixtures (KVA64/IbuNa/SA).

In contrast to other studies conducted on plasticization of polymers with di-carboxylic acids and their esters [31,32,63], this work does not show a clear link of plasticization to the reduction of the mechanical properties. For example, both succinic and maleic acid decreased the hardness by the same value with no significant difference despite having different plasticizing properties. Moreover, increasing the proportion of SA improved plasticization but did not deteriorate the hardness. Based on these observations, it can be concluded that the mechanical properties were mainly influenced by the proportion of copovidone in the powder mixtures. Indeed, a lower amount of Kollidon VA64 produces less sintered zones and more porosity [9,11], which could reduce the hardness of the SOFs.

Table 4 shows that the longer disintegration time was observed for the SOFs sintered only with KVA64 (73 s). The introduction of other elements in the formulation reduced the disintegration time significantly below 60 s. The lowest values were observed for KVA64/IbuAc, KVA64/MA, binary KVA64/SA mixtures and ternary KVA64/IbuNa/SA mixtures, and no statistically significant difference was observed. The reduction in disintegration time could be correlated to the lower proportion of copovidone, which decreases the viscosity of the medium and slows down polymer erosion [17]. Overall, the SOFs disintegrated within the 3 min, which makes them suitable as orally disintegrating printlets according to the European Pharmacopeia [41].

Figure 8 shows the dissolution profiles of the different SOFs containing IbuNa and IbuAc. All formulations achieved more than 85% dissolution at 15 min, making them suitable for immediate release [64]. The formulation KVA64 95%/IbuNa 5% as well as KVA64 95%/IbuAc 5% exhibited 100% drug release at 10 min. The introduction of succinic acid at a high percentage (10, 15, 20%) in the formulation of KVA64 and IbuNa, slowed down drug release. This could be due to the acidification of the SOF's microenvironment by the organic acid as previously mentioned by Sateesha et al. [65], which decreases the ionization of ibuprofen and hence its solubility.

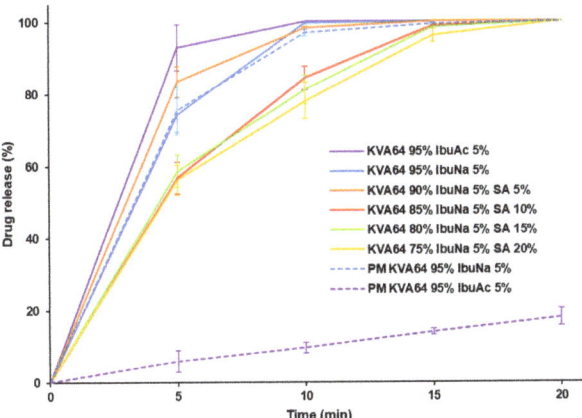

Figure 8. Dissolutions profiles of the different SOFs prepared with drugs and the physical mixtures of KVA64 95%/IbuNa 5% and KVA64 95%/IbuAc 5%.

Figure 8 exhibits the differences in drug release between the physical mixtures and the prepared SOFs for both KVA64 95%/IbuNa 5% and KVA64 95%/IbuAc 5%. It can be noted that while amorphization induced by SLS did not influence the solubility of IbuNa, which is already high prior to sintering, the solubility of IbuAc significantly improved. At 20 min, full dissolution of the SOF was achieved whereas the physical mixture of IbuAc/KVA64 dissolved only up to 18%. This confirms the ability of SLS to improve the solubility of poorly soluble drugs by preparing amorphous solid dispersions [65,66].

In general, the influence of di-carboxylic acids and ibuprofen acid on the SOF's properties could not be attributed to their plasticizing efficiency in SLS, as even non-plasticizing IbuNa and tartaric acid showed similar effects. Other factors were determinant for the properties of SOFs, such as the laser energy density, the compactness of the powder and the nature of the polymeric carrier. Amorphous copovidone already presents a low T_g compared to other polymers, which explains the low mechanical properties of the sintered SOFs [45]. Its low compactness is also involved in the reduction of the hardness [17]. Most importantly, its high solubility in water explains the rapid disintegration of the SOFs and the fast release of the drug [66].

3.5. Drug Degradation Evaluation

UHPLC analysis of the different SOFs printed with API revealed only one chromatographic peak corresponding to ibuprofen at the retention time (t_r = 1.50 min). In addition, the drug concentration in the different SOFs was analysed and it was found that the drug content was in agreement with the initial loading of the formulation (Table 5). This proves that no degradation has occurred during the sintering process, neither for the acid form nor for the sodium salt form. Therefore, no conclusion was possible regarding the presumed protective effect of SA as no degradation has been detected whether in absence or presence of the acid. As for IbuAc, it did not degrade despite its thermolability, suggesting a potential "autoprotective" effect due to its ability to decrease the optimal heating temperature. Further work will be needed using more thermolabile drugs without plasticizing properties in order to evidence the protective effect of succinic acid.

Table 5. Drug content of the different SOFs printed with drugs.

Formulation	Drug Content (%)
KVA64/IbuAc	102.2 ± 2.0
KVA64/IbuNa	96.5 ± 2.7
KVA64/IbuNa/SA	103.0 ± 4.1
KVA64/IbuNa/SA10	102.8 ± 3.6
KVA64/IbuNa/SA15	104.1 ± 4.5
KVA64/IbuNa/SA20	103.0 ± 2.2

3.6. Mechanisms of Plasticization in SLS

Although DSC and FTIR techniques have evidenced a decrease in the T_g and the presence of interactions between the components used in this study to sinter SOFs, they did not clearly discriminate between the different di-carboxylic acids as well as the two forms of ibuprofen. Nevertheless, hypotheses on the plasticization mechanisms can be formulated based on the chemical structure of the different excipients and drugs.

Among the used di-carboxylic acids, succinic acid exhibited the highest plasticizing effect and allowed the heating temperature to be reduced by 15 °C. This shows that the carboxylic groups played a role in plasticization, presumably by establishing H-bonds with the C=O functions of copovidone [52,67,68]. The plasticizing efficiency of succinic acid increased with its percentage in the formulation. However, above 15% the optimal heating temperature remained the same. These results can be explained by either a potential phase separation occurring at high content of plasticizer which prevents the creation of interactions with the polymer [69]. Or another reason could be that above 15%, an excess of H-bonds is formed between the acid and the polymer which would reduce the mobility of the polymer chains instead of increasing it [44]. Furthermore, renewable materials already produced on an industrial scale such as succinic acid seem good candidates to replace current synthetic plasticizers [70].

When more rigidity was introduced into the structure of the di-carboxylic acid (fumaric and maleic acids), less effect was observed in lowering the optimal heating temperature compared to succinic acid (5 °C instead of 15 °C). It seems that introducing a double bond in the structure provides less flexibility for the plasticizers to interact with the copovidone [71]. Furthermore, switching from the conformation *trans* (fumaric acid) to the conformation *cis* (maleic acid) did not have an impact in processability.

The effect of adding more hydroxyls into the structure while maintaining similar flexibility was studied by comparing succinic acid to malic acid and tartaric acid. In this case, the introduction of hydroxyl groups did not promote plasticization but instead reduced it, as previously seen with polyvinyl alcohol/starch films plasticized with polyols [46]. In this example, mannitol demonstrated inferior plasticizing properties than hexanetriol due to the presence of three additional hydroxyl groups in the structure, which increased its molecular weight and prevented its diffusion in the molecular matrix. Furthermore, the increased number of potential hydrogen bonds could also make the polymer structure more rigid and reduce the mobility of the chains [44]. This phenomenon is described as antiplasticization and is caused by a reduction of the free volume of the polymer by "filling the holes" with small molecules [72].

A noticeable effect on the processability was also observed when ibuprofen sodium salt was replaced by ibuprofen acid (a difference of 40 °C in OHT between KVA64 95%/IbuNa 5% and KVA64 95% / IbuAc 5%). A potential explanation for this difference in temperature is the formation of interactions in the forms prepared with copovidone and ibuprofen acid, as opposed to the salt form which do not exhibit interactions with the polymer [73]. Ibuprofen acid is known as a "non-traditional" plasticizer [39]. These non-traditional plasticizers are drugs that, in addition to improving the mechanical properties of films, could also provide a technical advantage by lowering the temperature during the hot-melt extrusion process [74]. As previously mentioned, IbuAc is poorly soluble in water and unstable at high temperature. However, during SLS its plasticizing effect may protect it from degradation and its potential amorphization during the sintering process can be an asset to enhance its solubility. As for IbuNa, replacing the hydrogen in the carboxylic function by sodium atoms blocks the plasticizing effect but enhances both solubility and thermal stability. Ibuprofen acid exhibited a higher plasticizing effect compared to succinic acid, despite its higher molecular weight and the presence of only one carboxylic group. This could be explained by the ability of IbuAc to establish aromatic bonds in addition to H-bonds with poly(vinylpyrrolidone) as previously reported by Bogdanova et al. [75]. This was observed in our previous work [17] with paracetamol, which also has an aromatic ring and acts as a plasticizer.

4. Conclusions

This study represents a proof of concept that di-carboxylic acids can be used as potential plasticizers to decrease the optimal heating temperature for the selective laser sintering of pharmaceutical solid oral forms prepared with copovidone. Depending on the chemical structure of the acid, a different effect on processability was demonstrated. Succinic acid was identified as the most performant plasticizer, as it allows a higher decrease in the optimal heating temperature when incorporated. Furthermore, the model drug used in this work, ibuprofen, in its acid form acts as a non-traditional plasticizer by lowering the heating temperature considerably, opposed to the sodium salt ibuprofen which exhibited no influence on the process.

DSC analysis confirmed the plasticizing effect of the different excipients and drugs by evidencing a decrease in a glass transition temperature and FTIR analysis showed the presence of interactions between the polymer and the di-carboxylic acids or the drugs in the sintered solid oral forms. Both techniques displayed limits when discriminating between the different plasticizers. However, based on the chemical structure of the components, it was concluded that interactions such as hydrogen bonds promote plasticization, but an excess could inversely reduce the mobility of polymer chains. In the future, other techniques could be employed to provide more insight on the mechanisms of plasticization, such as FTIR at variable temperature and molecular dynamic simulations.

Moreover, introduction of excipients and/or drugs in the formulation modified the properties (weight, hardness, disintegration time) of the solid oral forms, but this effect was more correlated to the decrease in the proportion of the polymer than the plasticization. Their beneficial effect on protecting the drug from degradation was not demonstrated since both forms of ibuprofen remained stable at the printing parameters. This suggests that these plasticizers aim more to facilitate the process than improving the properties of the printed solid oral forms.

Overall, this work highlights the importance of understanding the relationship between the material properties and the process parameters, especially with innovative technologies such as selective laser sintering. Further work is encouraged on this research path with different polymers and more thermosensitive drugs, in order to explore more profoundly the advantages of using di-carboxylic acids in SLS.

Supplementary Materials: The following are available online at https://www.mdpi.com/article/10.3390/polym13193282/s1, Figure S1: DSC thermograms of KVA64 during the 1st and 2nd heatings. Figure S2: DSC thermogram of IbuNa. Figure S3. Images of the mixtures KVA64 95%/SA 5% and KVA64 95%/TA 5%: (a) SEM images of the powders (prior to sintering) (Magnification $\times$ 100), (b) SEM images of the SOFs vertical sections (Magnification $\times$ 35), images of the SOFs. Table S1: Bulk density and true density of the different printed powders.

Author Contributions: Conceptualization, Y.A.G., N.M.S.-B. and I.S.; Data curation, Y.A.G.; Investigation, Y.A.G., N.M.S.-B., A.A. and J.-C.R.; Methodology, Y.A.G., N.M.S.-B. and I.S.; Software, A.A.; Supervision, B.B. and I.S.; Validation, N.M.S.-B., B.B. and I.S.; Visualization, N.M.S.-B., B.B. and I.S.; Writing—original draft, Y.A.G.; Writing—review and editing, N.M.S.-B., B.B. and I.S. All authors have read and agreed to the published version of the manuscript.

Funding: This research received no external funding.

Data Availability Statement: Data is contained within the article or supplementary material.

Acknowledgments: This work was supported by the funding program from the Algerian Ministry of High Education and Scientific Research through the funding of Yanis Abdelhamid GUECHE.

Conflicts of Interest: The authors declare no conflict of interest.

References

1. Elkasabgy, N.A.; Mahmoud, A.A.; Maged, A. 3D Printing: An Appealing Route for Customized Drug Delivery Systems. *Int. J. Pharm.* **2020**, *588*, 119732. [CrossRef]
2. Vaz, V.M.; Kumar, L. 3D Printing as a Promising Tool in Personalized Medicine. *AAPS PharmSciTech* **2021**, *22*, 49. [CrossRef]
3. Beg, S.; Almalki, W.H.; Malik, A.; Farhan, M.; Aatif, M.; Rahman, Z.; Alruwaili, N.K.; Alrobaian, M.; Tarique, M.; Rahman, M. 3D Printing for Drug Delivery and Biomedical Applications. *Drug Discov. Today* **2020**, *25*, 1668–1681. [CrossRef]
4. Trenfield, S.J.; Awad, A.; Madla, C.M.; Hatton, G.B.; Firth, J.; Goyanes, A.; Gaisford, S.; Basit, A.W. Shaping the Future: Recent Advances of 3D Printing in Drug Delivery and Healthcare. *Expert Opin. Drug Deliv.* **2019**, *16*, 1081–1094. [CrossRef] [PubMed]
5. Apprecia Pharmaceuticals What Is SPRITAM? Available online: https://www.spritam.com/#/patient/about-spritam/what-is-spritam (accessed on 7 October 2019).
6. Awad, A.; Fina, F.; Goyanes, A.; Gaisford, S.; Basit, A.W. 3D Printing: Principles and Pharmaceutical Applications of Selective Laser Sintering. *Int. J. Pharm.* **2020**, *586*, 119594. [CrossRef] [PubMed]
7. Charoo, N.A.; Barakh Ali, S.F.; Mohamed, E.M.; Kuttolamadom, M.A.; Ozkan, T.; Khan, M.A.; Rahman, Z. Selective Laser Sintering 3D Printing – an Overview of the Technology and Pharmaceutical Applications. *Drug Dev. Ind. Pharm.* **2020**, *46*, 869–877. [CrossRef] [PubMed]
8. Fina, F.; Madla, C.M.; Goyanes, A.; Zhang, J.; Gaisford, S.; Basit, A.W. Fabricating 3D Printed Orally Disintegrating Printlets Using Selective Laser Sintering. *Int. J. Pharm.* **2018**, *541*, 101–107. [CrossRef] [PubMed]
9. Allahham, N.; Fina, F.; Marcuta, C.; Kraschew, L.; Mohr, W.; Gaisford, S.; Basit, A.W.; Goyanes, A. Selective Laser Sintering 3D Printing of Orally Disintegrating Printlets Containing Ondansetron. *Pharmaceutics* **2020**, *12*, 110. [CrossRef]
10. Fina, F.; Goyanes, A.; Madla, C.M.; Awad, A.; Trenfield, S.J.; Kuek, J.M.; Patel, P.; Gaisford, S.; Basit, A.W. 3D Printing of Drug-Loaded Gyroid Lattices Using Selective Laser Sintering. *Int. J. Pharm.* **2018**, *547*, 44–52. [CrossRef]
11. Barakh Ali, S.F.; Mohamed, E.M.; Ozkan, T.; Kuttolamadom, M.A.; Khan, M.A.; Asadi, A.; Rahman, Z. Understanding the Effects of Formulation and Process Variables on the Printlets Quality Manufactured by Selective Laser Sintering 3D Printing. *Int. J. Pharm.* **2019**, *570*, 118651. [CrossRef] [PubMed]
12. Mohamed, E.M.; Barakh Ali, S.F.; Rahman, Z.; Dharani, S.; Ozkan, T.; Kuttolamadom, M.A.; Khan, M.A. Formulation Optimization of Selective Laser Sintering 3D-Printed Tablets of Clindamycin Palmitate Hydrochloride by Response Surface Methodology. *AAPS PharmSciTech* **2020**, *21*, 232. [CrossRef]
13. Aho, J.; Bøtker, J.P.; Genina, N.; Edinger, M.; Arnfast, L.; Rantanen, J. Roadmap to 3D-Printed Oral Pharmaceutical Dosage Forms: Feedstock Filament Properties and Characterization for Fused Deposition Modeling. *J. Pharm. Sci.* **2019**, *108*, 26–35. [CrossRef]
14. Yang, Y.; Xu, Y.; Wei, S.; Shan, W. Oral Preparations with Tunable Dissolution Behavior Based on Selective Laser Sintering Technique. *Int. J. Pharm.* **2020**, 120127. [CrossRef]
15. Goodridge, R.D.; Tuck, C.J.; Hague, R.J.M. Laser Sintering of Polyamides and Other Polymers. *Prog. Mater. Sci.* **2012**, *57*, 229–267. [CrossRef]
16. Chatham, C.A.; Long, T.E.; Williams, C.B. A Review of the Process Physics and Material Screening Methods for Polymer Powder Bed Fusion Additive Manufacturing. *Prog. Polym. Sci.* **2019**, *93*, 68–95. [CrossRef]
17. Gueche, Y.A.; Sanchez-Ballester, N.M.; Bataille, B.; Aubert, A.; Leclercq, L.; Rossi, J.-C.; Soulairol, I. Selective Laser Sintering of Solid Oral Dosage Forms with Copovidone and Paracetamol Using a CO_2 Laser. *Pharmaceutics* **2021**, *13*, 160. [CrossRef] [PubMed]
18. Zhao, Y.; Xie, X.; Zhao, Y.; Gao, Y.; Cai, C.; Zhang, Q.; Ding, Z.; Fan, Z.; Zhang, H.; Liu, M.; et al. Effect of Plasticizers on Manufacturing Ritonavir/Copovidone Solid Dispersions via Hot-Melt Extrusion: Preformulation, Physicochemical Characterization, and Pharmacokinetics in Rats. *Eur. J. Pharm. Sci.* **2019**, *127*, 60–70. [CrossRef] [PubMed]
19. Cailleaux, S.; Sanchez-Ballester, N.M.; Gueche, Y.A.; Bataille, B.; Soulairol, I. Fused Deposition Modeling (FDM), the New Asset for the Production of Tailored Medicines. *J. Contro. Release* **2020**, S0168365920306374. [CrossRef]
20. Lexow, M.M.; Drexler, M.; Drummer, D. Fundamental Investigation of Part Properties at Accelerated Beam Speeds in the Selective Laser Sintering Process. *RPJ* **2017**, *23*, 1099–1106. [CrossRef]
21. Schmid, M.; Amado, A.; Wegener, K. Materials Perspective of Polymers for Additive Manufacturing with Selective Laser Sintering. *J. Mater. Res.* **2014**, *29*, 1824–1832. [CrossRef]
22. Braun, D. *Polymer Synthesis: Theory and Practice: Fundamentals, Methods, Experiments*, 5th ed.; Springer: Berlin/Heidelberg, Germany; New York, NY, USA, 2013; ISBN 978-3-642-28979-8.
23. Mohsin, M.; Hossin, A.; Haik, Y. Thermomechanical Properties of Poly(Vinyl Alcohol) Plasticized with Varying Ratios of Sorbitol. *Mater. Sci. Eng. A* **2011**, *528*, 925–930. [CrossRef]
24. Jamarani, R.; Erythropel, H.; Nicell, J.; Leask, R.; Marić, M. How Green Is Your Plasticizer? *Polymers* **2018**, *10*, 834. [CrossRef] [PubMed]
25. Mayer, F.L.; Stalling, D.L.; Johnson, J.L. Phthalate Esters as Environmental Contaminants. *Nature* **1972**, *238*, 411–413. [CrossRef]

26. Fan, J.; Traore, K.; Li, W.; Amri, H.; Huang, H.; Wu, C.; Chen, H.; Zirkin, B.; Papadopoulos, V. Molecular Mechanisms Mediating the Effect of Mono-(2-Ethylhexyl) Phthalate on Hormone-Stimulated Steroidogenesis in MA-10 Mouse Tumor Leydig Cells. *Endocrinology* **2010**, *151*, 3348–3362. [CrossRef]
27. Debabov, V.G. Prospects for Biosuccinic Acid Production. *Appl. Biochem. Microbiol.* **2015**, *51*, 787–791. [CrossRef]
28. Sebastian, J.; Hegde, K.; Kumar, P.; Rouissi, T.; Brar, S.K. Bioproduction of Fumaric Acid: An Insight into Microbial Strain Improvement Strategies. *Crit. Rev. Biotechnol.* **2019**, *39*, 817–834. [CrossRef] [PubMed]
29. Howell, B.A.; Sun, W. Biobased Plasticizers from Tartaric Acid, an Abundantly Available, Renewable Material. *Ind. Eng. Chem. Res.* **2018**, acs.iecr.8b03486. [CrossRef]
30. Llanes, L.C. Mechanical and Thermal Properties of Poly(Lactic Acid) Plasticized with Dibutyl Maleate and Fumarate Isomers: Promising Alternatives as Biodegradable Plasticizers. *Eur. Polym. J.* **2021**, *142*, 110112. [CrossRef]
31. Stuart, A.; McCallum, M.M.; Fan, D.; LeCaptain, D.J.; Lee, C.Y.; Mohanty, D.K. Poly(Vinyl Chloride) Plasticized with Succinate Esters: Synthesis and Characterization. *Polym. Bull.* **2010**, *65*, 589–598. [CrossRef]
32. Ahmed, I.; Niazi, M.B.K.; Hussain, A.; Jahan, Z. Influence of Amphiphilic Plasticizer on Properties of Thermoplastic Starch Films. *Polym.-Plast. Technol. Eng.* **2018**, *57*, 17–27. [CrossRef]
33. Daniels, P.H. A Brief Overview of Theories of PVC Plasticization and Methods Used to Evaluate PVC-Plasticizer Interaction. *J. Vinyl. Addit. Technol.* **2009**, *15*, 219–223. [CrossRef]
34. Kirkpatrick, A. Some Relations Between Molecular Structure and Plasticizing Effect. *J. Appl. Phys.* **1940**, *11*, 255–261. [CrossRef]
35. Aiken, W.; Alfrey, T.; Janssen, A.; Mark, H. Creep Behavior of Plasticized Vinylite VYNW. *J. Polym. Sci.* **1947**, *2*, 178–198. [CrossRef]
36. Spurlin, H.M. The Technology of Solvents and Plasticizers. *J. Polym. Sci.* **1955**, *18*, 444–445. [CrossRef]
37. Williams, M.L.; Landel, R.F.; Ferry, J.D. The Temperature Dependence of Relaxation Mechanisms in Amorphous Polymers and Other Glass-Forming Liquids. *J. Am. Chem. Soc.* **1955**, *77*, 3701–3707. [CrossRef]
38. Kolter, L.; Karl, M.; Gryczke, A. *Hot-Melt Extrusion with BASF Pharma Polymers. Extrusion Compendium*, 2nd ed.; BASF: Ludwigshafen, Germany, 2012.
39. Wu, C.; McGinity, J.W. Non-Traditional Plasticization of Polymeric Films. *Int. J. Pharm.* **1999**, *177*, 15–27. [CrossRef]
40. Caviglioli, G.; Valeria, P.; Brunella, P.; Sergio, C.; Attilia, A.; Gaetano, B. Identification of Degradation Products of Ibuprofen Arising from Oxidative and Thermal Treatments. *J. Pharm. Biomed. Anal.* **2002**, *30*, 499–509. [CrossRef]
41. European Directorate for the Quality of Medicines & Healthcare. Council of Europe Disintegration of tablets and capsules (monograph 2.9.1). In *European Pharmacopeia*; Council of Europe: Strasbourg, France, 2019; pp. 323–325.
42. European Directorate for the Quality of Medicines & Healthcare. Council of Europe Dissolution test for solid dosage forms (monograph 2.9.3). In *European Pharmacopeia*; Council of Europe: Strasbourg, France, 2019; pp. 326–333.
43. Surana, R.; Randall, L.; Pyne, A.; Vemuri, N.M.; Suryanarayanan, R. Determination of Glass Transition Temperature and in Situ Study of the Plasticizing Effect of Water by Inverse Gas Chromatography. *Pharm. Res.* **2003**, *20*, 1647–1654. [CrossRef]
44. Özeren, H.D.; Guivier, M.; Olsson, R.T.; Nilsson, F.; Hedenqvist, M.S. Ranking Plasticizers for Polymers with Atomistic Simulations: PVT, Mechanical Properties, and the Role of Hydrogen Bonding in Thermoplastic Starch. *ACS Appl. Polym. Mater.* **2020**, *2*, 2016–2026. [CrossRef]
45. Dickens, E.D.; Lee, B.L.; Taylor, G.A.; Magistro, A.J.; Ng, H.; McAlea, K.; Forderhase, P.F. Sinterable Semi-Crystalline Powder and near Fully Dense Article Formed Therewith 2000. U.S. Patent US5527877A, 18 June 1996.
46. Aydın, A.A.; Ilberg, V. Effect of Different Polyol-Based Plasticizers on Thermal Properties of Polyvinyl Alcohol:Starch Blends. *Carbohydr. Polym.* **2016**, *136*, 441–448. [CrossRef]
47. Sun, S.; Song, Y.; Zheng, Q. Thermo-Molded Wheat Gluten Plastics Plasticized with Glycerol: Effect of Molding Temperature. *Food Hydrocoll.* **2008**, *22*, 1006–1013. [CrossRef]
48. Ligon, S.C.; Liska, R.; Stampfl, J.; Gurr, M.; Mülhaupt, R. Polymers for 3D Printing and Customized Additive Manufacturing. *Chem. Rev.* **2017**, *117*, 10212–10290. [CrossRef] [PubMed]
49. Censi, R.; Martena, V.; Hoti, E.; Malaj, L.; Di Martino, P. Sodium Ibuprofen Dihydrate and Anhydrous: Study of the Dehydration and Hydration Mechanisms. *J. Anal. Calorim.* **2013**, *111*, 2009–2018. [CrossRef]
50. Ma, X.; Qiao, C.; Wang, X.; Yao, J.; Xu, J. Structural Characterization and Properties of Polyols Plasticized Chitosan Films. *Int. J. Biol. Macromol.* **2019**, *135*, 240–245. [CrossRef] [PubMed]
51. Pereira, B.C.; Isreb, A.; Forbes, R.T.; Dores, F.; Habashy, R.; Petit, J.-B.; Alhnan, M.A.; Oga, E.F. 'Temporary Plasticiser': A Novel Solution to Fabricate 3D Printed Patient-Centred Cardiovascular 'Polypill' Architectures. *Eur. J. Pharm. Biopharm.* **2019**, *135*, 94–103. [CrossRef]
52. Hurley, D.; Carter, D.; Foong Ng, L.Y.; Davis, M.; Walker, G.M.; Lyons, J.G.; Higginbotham, C.L. An Investigation of the Inter-Molecular Interaction, Solid-State Properties and Dissolution Properties of Mixed Copovidone Hot-Melt Extruded Solid Dispersions. *J. Drug Deliv. Sci. Technol.* **2019**, *53*, 101132. [CrossRef]
53. Yuan, X.; Xiang, T.-X.; Anderson, B.D.; Munson, E.J. Hydrogen Bonding Interactions in Amorphous Indomethacin and Its Amorphous Solid Dispersions with Poly(Vinylpyrrolidone) and Poly(Vinylpyrrolidone-Co-Vinyl Acetate) Studied Using 13C Solid-State NMR. *Mol. Pharm.* **2015**, *12*, 4518–4528. [CrossRef]
54. Zhang, Y.; Grant, D.J.W. Similarity in Structures of Racemic and Enantiomeric Ibuprofen Sodium Dihydrates. *Acta Cryst. C Cryst. Struct. Commun.* **2005**, *61*, m435–m438. [CrossRef]

55. Matet, M.; Heuzey, M.-C.; Pollet, E.; Ajji, A.; Avérous, L. Innovative Thermoplastic Chitosan Obtained by Thermo-Mechanical Mixing with Polyol Plasticizers. *Carbohydr. Polym.* **2013**, *95*, 241–251. [CrossRef]
56. Maddineni, S.; Battu, S.K.; Morott, J.; Majumdar, S.; Murthy, S.N.; Repka, M.A. Influence of Process and Formulation Parameters on Dissolution and Stability Characteristics of Kollidon®VA 64 Hot-Melt Extrudates. *AAPS PharmSciTech* **2015**, *16*, 444–454. [CrossRef]
57. Schramm, C. High Temperature ATR-FTIR Characterization of the Interaction of Polycarboxylic Acids and Organotrialkoxysilanes with Cellulosic Material. *Spectrochim. Acta Part A Mol. Biomol. Spectrosc.* **2020**, *243*, 118815. [CrossRef]
58. Choperena, A.; Painter, P. Hydrogen Bonding in Polymers: Effect of Temperature on the OH Stretching Bands of Poly(Vinylphenol). *Macromolecules* **2009**, *42*, 6159–6165. [CrossRef]
59. Lee, T.; Wang, Y.W. Initial Salt Screening Procedures for Manufacturing Ibuprofen. *Drug Dev. Ind. Pharm.* **2009**, *35*, 555–567. [CrossRef] [PubMed]
60. Davis, S.S.; Hardy, J.G.; Taylor, M.J.; Whalley, D.R.; Wilson, C.G. A Comparative Study of the Gastrointestinal Transit of a Pellet and Tablet Formulation. *Int. J. Pharm.* **1984**, *21*, 167–177. [CrossRef]
61. Hamed, R.; Mohamed, E.M.; Rahman, Z.; Khan, M.A. 3D-Printing of Lopinavir Printlets by Selective Laser Sintering and Quantification of Crystalline Fraction by XRPD-Chemometric Models. *Int. J. Pharm.* **2020**, 120059. [CrossRef]
62. Yu, Q.; Dang, L.; Black, S.; Wei, H. Crystallization of the Polymorphs of Succinic Acid via Sublimation at Different Temperatures in the Presence or Absence of Water and Isopropanol Vapor. *J. Cryst. Growth* **2012**, *340*, 209–215. [CrossRef]
63. Enumo, A.; Gross, I.P.; Saatkamp, R.H.; Pires, A.T.N.; Parize, A.L. Evaluation of Mechanical, Thermal and Morphological Properties of PLA Films Plasticized with Maleic Acid and Its Propyl Ester Derivatives. *Polym. Test.* **2020**, *88*, 106552. [CrossRef]
64. Committee for Medicinal Products for Human Use. *Guideline on the Investigation of Bioequivalence*; European Medicines Agency: Amsterdam, The Netherlands, 2010.
65. Sateesha, S.B.; Narode, M.K.; Vyas, B.D.; Rajamma, A.J. Influence of Organic Acids on Diltiazem HCl Release Kinetics from Hydroxypropyl Methyl Cellulose Matrix Tablets. *J. Young Pharm.* **2010**, *2*, 229–233. [CrossRef]
66. Kollamaram, G.; Croker, D.M.; Walker, G.M.; Goyanes, A.; Basit, A.W.; Gaisford, S. Low Temperature Fused Deposition Modeling (FDM) 3D Printing of Thermolabile Drugs. *Int. J. Pharm.* **2018**, *545*, 144–152. [CrossRef]
67. Bellelli, M.; Licciardello, F.; Pulvirenti, A.; Fava, P. Properties of Poly(Vinyl Alcohol) Films as Determined by Thermal Curing and Addition of Polyfunctional Organic Acids. *Food Packag. Shelf Life* **2018**, *18*, 95–100. [CrossRef]
68. Park, H.-R.; Chough, S.-H.; Yun, Y.-H.; Yoon, S.-D. Properties of Starch/PVA Blend Films Containing Citric Acid as Additive. *J. Polym. Environ.* **2005**, *13*, 375–382. [CrossRef]
69. Talja, R.A.; Helén, H.; Roos, Y.H.; Jouppila, K. Effect of Various Polyols and Polyol Contents on Physical and Mechanical Properties of Potato Starch-Based Films. *Carbohydr. Polym.* **2007**, *67*, 288–295. [CrossRef]
70. Mazière, A.; Prinsen, P.; García, A.; Luque, R.; Len, C. A Review of Progress in (Bio)Catalytic Routes from/to Renewable Succinic Acid. *Biofuels Bioprod. Bioref.* **2017**, *11*, 908–931. [CrossRef]
71. Streitwieser, A.; Heathcock, C.H.; Kosower, E.M. *Introduction to Organic Chemistry*, 4th ed.; Macmillan: New York, NY, USA; Maxwell Macmillan Canada: Toronto, ON, Canada; Maxwell Macmillan International: New York, NY, USA, 1992; ISBN 978-0-02-418170-1.
72. Vrentas, J.S.; Duda, J.L.; Ling, H.C. Antiplasticization and Volumetric Behavior in Glassy Polymers. *Macromolecules* **1988**, *21*, 1470–1475. [CrossRef]
73. Pudlas, M.; Kyeremateng, S.O.; Williams, L.A.M.; Kimber, J.A.; van Lishaut, H.; Kazarian, S.G.; Woehrle, G.H. Analyzing the Impact of Different Excipients on Drug Release Behavior in Hot-Melt Extrusion Formulations Using FTIR Spectroscopic Imaging. *Eur. J. Pharm. Sci.* **2015**, *67*, 21–31. [CrossRef]
74. Repka, M.A.; McGinity, J.W. Influence of Vitamin E TPGS on the Properties of Hydrophilic Films Produced by Hot-Melt Extrusion. *Int. J. Pharm.* **2000**, *202*, 63–70. [CrossRef]
75. Bogdanova, S.; Pajeva, I.; Nikolova, P.; Tsakovska, I.; Müller, B. Interactions of Poly(Vinylpyrrolidone) with Ibuprofen and Naproxen: Experimental and Modeling Studies. *Pharm. Res.* **2005**, *22*, 806–815. [CrossRef] [PubMed]

Article

Low Temperature Powder Bed Fusion of Polymers by Means of Fractal Quasi-Simultaneous Exposure Strategies

Samuel Schlicht [1,*], Sandra Greiner [1,2] and Dietmar Drummer [1]

1. Institute of Polymer Technology, Friedrich-Alexander-Universität Erlangen-Nürnberg, Am Weichselgarten 10, 91058 Erlangen, Germany; sandra.greiner@fau.de (S.G.); dietmar.drummer@fau.de (D.D.)
2. Collaborative Research Center 814, Friedrich-Alexander-Universität Erlangen-Nürnberg, Am Weichselgarten 10, 91058 Erlangen, Germany
* Correspondence: samuel.schlicht@fau.de

Abstract: Powder Bed Fusion of Polymers (PBF-LB/P) is a layer-wise additive manufacturing process that predominantly relies on the quasi-isothermal processing of semi-crystalline polymers, inherently limiting the spectrum of polymers suitable for quasi-isothermal PBF. Within the present paper, a novel approach for extending the isothermal processing window towards significantly lower temperatures by applying the quasi-simultaneous laser-based exposure of fractal scan paths is proposed. The proposed approach is based on the temporal and spatial discretization of the melting and subsequent crystallization of semi-crystalline thermoplastics, hence allowing for the mesoscale compensation of crystallization shrinkage of distinct segments. Using thermographic monitoring, a homogenous temperature increase of discrete exposed sub-segments, limited thermal interference of distinct segments, and the resulting avoidance of curling and warping can be observed. Manufactured parts exhibit a dense and lamellar part morphology with a nano-scale semi-crystalline structure. The presented approach represents a novel methodology that allows for significantly reducing energy consumption, process preparation times and temperature-induced material aging in PBF-LB/P while representing the foundation for the processing of novel, thermo-sensitive material systems in PBF-LB/P.

Keywords: powder bed fusion; laser sintering; isothermal; low temperature laser sintering; selective laser melting

Citation: Schlicht, S.; Greiner, S.; Drummer, D. Low Temperature Powder Bed Fusion of Polymers by Means of Fractal Quasi-Simultaneous Exposure Strategies. *Polymers* **2022**, *14*, 1428. https://doi.org/10.3390/polym14071428

Academic Editors: Swee Leong Sing and Wai Yee Yeong

Received: 21 February 2022
Accepted: 28 March 2022
Published: 31 March 2022

Publisher's Note: MDPI stays neutral with regard to jurisdictional claims in published maps and institutional affiliations.

Copyright: © 2022 by the authors. Licensee MDPI, Basel, Switzerland. This article is an open access article distributed under the terms and conditions of the Creative Commons Attribution (CC BY) license (https://creativecommons.org/licenses/by/4.0/).

1. Introduction

Laser-based powder bed fusion of polymers (PBF-LB/P) is a powder-based additive manufacturing process that allows for manufacturing individualized components with a high geometric freedom. To date, PBF-LB/P is predominantly associated with the quasi-isothermal processing of semi-crystalline polymers. Given the continuous heating of the build chamber in quasi-isothermal PBF-LB/P, temperature-induced aging of polymers [1], increased process times due to heating and cooling phases, and the influence of processing times on resulting mechanical properties [2,3] inherently restrict the economic and ecological viability for the cost-efficient manufacturing of polymer components. Considering non-uniform temperature fields occurring in PBF-LB/P, the isothermal assumption merely represents an idealization. With regard to isothermal crystallization kinetics of Polyamide 12, findings by Neugebauer et al. [4] indicate the occurrence of isothermal crystallization during isothermal PBF of polymers. Using a process-integrated approach, Drummer et al. [5] determined the time-dependent occurrence of isothermal crystallization in PBF-LB/P, proposing the possibility of novel process strategies for limiting the isothermal processing zone to the powder bed surface. Considering a time- and temperature-dependency of the isothermal crystallization process, findings by Soldner et al. [6] indicate a non-uniform, geometry-dependent isothermal crystallization [7]. Findings derived by Shen et al. [8] based on a numerical approach indicate a correlation of the underlying cooling rate and resulting residual stress, thus affecting part distortion. Consequently, controlled isothermal crystallization is omnipresent in PBF-LB/P, being influenced by time-, temperature-, and geometry-dependent effects. However, even considering the occurrence of isothermal crystallization in PBF-LB/P, the non-isothermal processing below the crystallization onset temperature of semi-crystalline polymers remains restricted due to the occurrence of stress-induced distortion, specifically curling and warping [9]. Induced by the inhomogeneous crystallization of the polymer melt, curling considerably reduces the process stability, leading to process interruptions. Crystallization-induced shrinkage is inherently bound to the processing of semi-crystalline polymers, thus constituting the requirement of novel strategies for controlling the crystallization kinetics for promoting a uniform, controlled crystallization of each layer. Therefore, the non-uniform crystallization of the applied materials constitutes an inherent challenge for novel processing strategies, focusing on the exposure-induced process optimization.

2. State of the Art
2.1. Kinetics of Isothermal and Non-Isothermal Crystallization of Polymers

Quasi-isothermal processing composes the state of the art in laser-based powder bed fusion of polymers. The isothermal assumption in PBF-LB/P implies the predominant occurrence of isothermal crystallization during the build process. A basic modelling of isothermal crystallization processes can be derived using the Avrami equation. Considering non-constant cooling rates, occurring in laser sintering of polymers in quasi-isothermal as well in non-isothermal processing, the Nakamura model, proposed by Nakamura et al. [10,11] allows for considering non-isothermal crystallization. The macroscopic degree of crystallization, α, can be expressed in dependence of the Nakamura kinetics crystallization function $K(T)$, being closely related to the Avrami function $k(T)$.

$$\alpha = 1 - \exp\left[-\left(\int_0^t K(t)\mathrm{d}t\right)^n\right] \quad (1)$$

The underlying relation of the Nakamura crystallization rate $K(t)$ and the Avrami crystallization rate $k(t)$ can be expressed using a temperature-dependent, dimensionless parameter n.

$$K(t) = k(t)^{\frac{1}{n}} \quad (2)$$

Ziabicki [12] described an empirical exponential relation of the crystallization half time $t_{1/2}$, the growth constant K_0 and the nucleation rate constant K_g by applying the Lauritzen–Hoffman theory, with the activation energy for polymer diffusion $U^* = 6270$ J mol^{-1}, the universal gas constant R, the temperature value $T_\infty = T_g - 30$ K, indicating a ceased viscous flow, and the equilibrium melting temperature T_m^0, displayed in Equation (3).

$$\frac{1}{t_{1/2}} = K_0 \exp\left[-\frac{U^*}{R(T-T_\infty)}\right] \exp\left[\frac{K_g(T+T_m^0)}{2T^2(T_m^0 - T)}\right] \quad (3)$$

Zhao et al. [13] applied the empirical relation on Polyamide 12, used in laser-based powder bed fusion, by fitting the parameters K_0 and K_g based on experimental data obtained from differential scanning calorimetry. The resulting crystallization half times, presented by Zhao et al., exhibit a satisfactory accordance of experimentally obtained and modelled values for sufficiently high cooling rates. Consequently, a reduced processing temperature T is correlated with considerably reduced crystallization half times. Considering isothermal crystallization of quenched Polyamide 12 at varying ambient temperatures, Paolucci et al. [14] modelled the temperature-dependent formation of varying crystalline phase compositions of Polyamide 12. Unpressurized crystallization at temperatures exceeding 100 °C predominantly leads to the formation of the α-phase [14], implying the formation of an identical crystalline phase within a wide thermal processing window.

However, with regard to described crystallization kinetics, a dependency of the applied cooling rate and morphological properties needs to be considered a major influence in low temperature PBF-LB/P.

2.2. Low Temperature Laser-Based Processing of Polymers

In contrast to powder bed fusion of metal alloys, PBF of polymers is characterized by the avoidance of support structures. With regard to the application of support structures, different approaches for transferring metal-based concepts on the non-isothermal processing of polymers [15] have been described. Resulting material morphologies exhibit significantly reduced spherulite sizes, correlated with an increased elongation at break while exhibiting an insufficient porosity level [16]. In contrast to the application of support structures, the avoidance of curling and warping by means of laser-based preheating has been proposed by Laumer et al. [17], applying simultaneous laser beam irradiation for the isothermal manufacturing of multi-material components. Investigations conducted by Chatham et al. [18] for the manufacturing of polyphenylene sulfide at reduced powder bed temperatures exhibit significantly increased levels of porosity, thus limiting the applicability of produced parts. Consequently, to date, the non-isothermal PBF of polymers is inherently limited with regard to the requirement of support structures and the emergence of insufficient part properties.

2.3. Influence of Exposure Strategies on Superficial Temperature Fields and Part Properties

The application of a variety of exposure strategies has been described for both metal-based and polymer-based PBF. The interaction of an exposed geometry and the applied exposure strategy on resulting temperature fields is described extensively in recent literature. Exhibiting a reduced thermal penetration depth, increased exposure speeds are correlated with increased superficial maximum temperature values [19,20]. In addition, resulting superficial temperature fields show a dependence on both the applied exposure speed and the underlying scan vector length [19], resulting from a thermal superposition of subsequently exposed scan vectors [20]. Jain et al. [2] describe a correlation of geometry-induced, varying return times of the laser beam and varying mechanical properties of samples manufactured using Polyamide 12, implying an influence of superficial temperature fields on resulting mechanical part properties. Exceeding the monitoring of exposure-induced temperature fields, Greiner et al. (2021) [21] observed an interdependence of applied exposure parameters and the underlying geometry on post-exposure temperature fields, leading to a varying morphological structure of fabricated parts.

Segmented exposure strategies, widely applied in laser-based PBF of metal alloys (PBF-LB/M), exhibit reduced residual stresses, described for the application of steel [22–24] and nickel alloys [25]. Zou et al. (2020) [22] describe a significant influence of the exposure sequence and orientation of distinct segments on resulting residual stress, emphasizing structural advantages of non-linear sequencing compared to linear sequencing. Considering varying exposure patterns of distinct segments, further complex interdependencies of the applied sequence of exposed scan vectors, the scan vector length and the sequence of exposed sub-segments are described. With regard to the inherent geometry-dependence of linear exposure patterns, non-linear exposure patterns are gaining increased attention both in metal- and polymer-based PBF. The application of non-linear exposure patterns was initially described by Yang et al. [26], applying the space-filling, fractal Hilbert curve for the sintering of polymer-bound ceramic particles. Further research on fractal exposure patterns was conducted by Ma et al. [27] and Catchpole-Smith et al. [28], describing reduced stress-induced distortion and the reduced occurrence of heat-induced cracks in PBF of nickel alloys. Greiner et al. [29] described the application of the fractal, space-filling Peano curve for the PBF of Polyamide 12, leading to geometry-invariant temperature fields promoted by the scale-invariant structure of the applied exposure pattern. Therefore, the application of linear exposure patterns is considerably influenced by the exposed cross-section, leading to a reduced reproducibility of part properties. In contrast, applying segmented, fractal exposure strategies promote the formation of uniform, geometry-invariant temperature fields that could be exploited for low temperature PBF-LB/P. Consequently, the formation of crystallization-induced residual stress and resulting part deformations exhibits a dependence of applied exposure strategies, hence implying the requirement of novel exposure strategies to overcome existing limitations of quasi-isothermal PBF of polymers.

3. Methodological Approach for Low Temperature PBF

The approach presented in this paper focusses on significantly lowering the build chamber temperature while limiting warping and curling of manufactured parts by means of fractal, quasi-simultaneous laser exposure. In contrast to quasi-isothermal PBF, low temperature PBF, as proposed in this paper, relies on the immediate crystallization of distinct exposed segments, considering a material-specific processing windows below the crystallization peak, displayed in Figure 1.

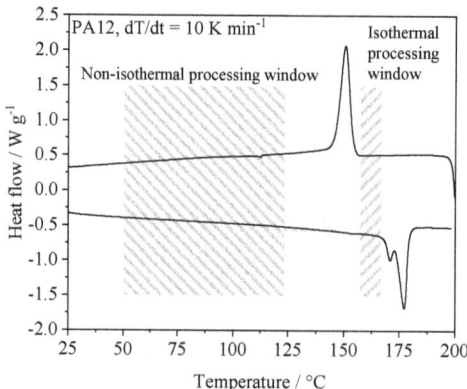

Figure 1. Schematic illustration of process-dependent thermal processing windows.

Resulting implications for process temperature control include significantly reduced pre-heating times and the immediate removal of manufactured parts subsequent to the build process. Reduced pre-heating times are obtained considering reduced requirements of the thermal homogeneity in contrast to quasi-isothermal processing, thus limiting the required homogenous thermal field to the thickness of the manufactured layer. Resulting process times of non-isothermal and quasi-isothermal processing, respectively, are displayed in Figure 2.

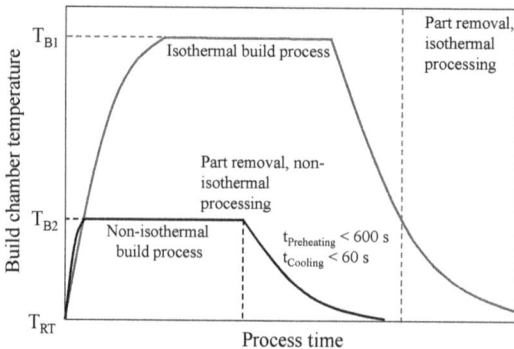

Figure 2. Schematic time-dependent temperature variation for applying quasi-isothermal and non-isothermal processing.

To allow for the non-isothermal processing of semi-crystalline polymers, restricting the distortion of exposed segments is essential. Based on previous research on the field of segmented exposure strategies [23,25], segmented exposure strategies are combined with the application of fractal scan paths [26–29] and quasi-simultaneous exposure of distinct segments. Fractal scan path generation applied within the present paper is based on space-filling, self-avoiding and self-similar curves, commonly referred to as "FASS curves" [30], specifically on the Peano curve [31]. The implementation of fractal, quasi-simultaneous exposure strategies is based on functional recursive programming, conducted in Python 3.8. Resulting exposure strategies are transferred using the Common Layer Interface (CLI) format to allow for the integration of complex exposure strategies into commercially available machinery. Exposure paths, applied for the non-isothermal processing of polymers, include discrete fractal sub-segments that are exposed using fractal sequencing. The resulting exposure strategy of an exemplary square cross-section is displayed in Figure 3.

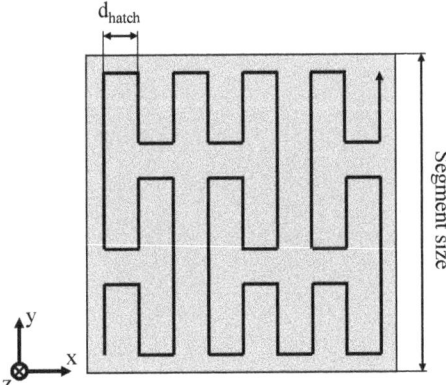

Figure 3. Schematic depiction of the applied fractal exposure pattern.

Quasi-simultaneous exposure of varying geometries is based on the repetitive, consecutive exposure of distinct fractal patterns. Each sub-segment constitutes a closed loop, allowing for an uninterrupted quasi-simultaneous exposure, schematically displayed in Figure 4.

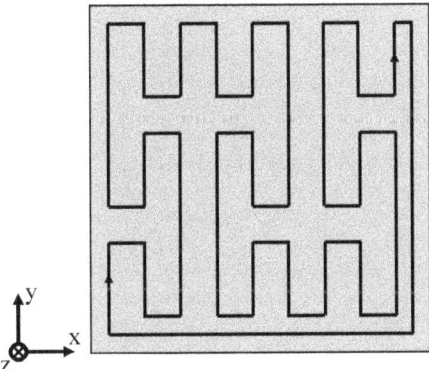

Figure 4. Fractal, quasi-simultaneous exposure pattern of a distinct sub-segment based on the Peano curve.

Quasi-simultaneous exposure is correlated with a significantly increased layer time due to the repetitive exposure of distinct segments, leading to an increase in the layer time equivalent to the additional number of exposure steps compared to single exposure. Distinct, repetitively exposed sub-segments are sequenced by applying a fractal exposure sequence to reduce geometry-induced influences and interferences on resulting temperature fields. Therefore, fractal patterns are applied on sub-segment level and for determining the sequence of consecutive segments. With regard to the scale-invariance of fractal space-filling curves, the sequence of consecutively scanned segments is determined by the structure of the Peano curve, schematically displayed in Figure 5.

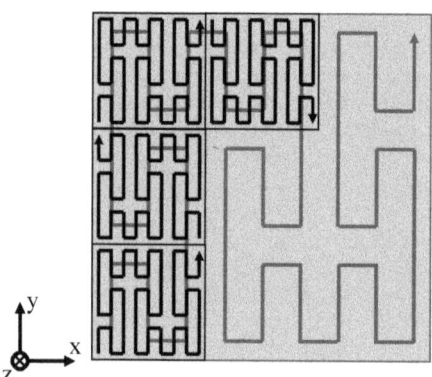

Figure 5. Schematic illustration of the fractal exposure sequence of distinct segments.

Exposure patterns of complex cross-sections are generated applying the algorithm proposed by Yang et al. [26], thus implicitly generating a closed, space-filling curve corresponding to the specific part contour.

4. Materials and Methods

4.1. Experimental Set-Up

All experimental work is conducted using a freely configurable SLS research system, thus allowing for the integration of complex space-filling scan paths, prepared by means of iterative algorithms. The used SLS machine resembles commercially available machinery, equipped with a high-speed galvanometer scanner, SCANLAB GmbH, Puchheim, Germany, while allowing for the direct control of the exposure sequence based on the programming of exposure paths. Constant optical parameters include a laser wavelength of $\lambda = 10.6$ µm and a laser focus diameter of $d = 0.5$ mm. A layer height of 0.1 mm is kept constant for all experimental investigations. Commercially available Polyamide 12 powder, PA 2200, EOS GmbH, Krailling, Germany, is used as the underlying material. A mixture of equal fractions of pre-used powder and virgin powder is applied. The pre-used powder was extracted from overflow bins and was exposed to a single recoating step, corresponding to merely insignificant thermal aging. Resulting material properties exhibit a bulk density of 0.44 g cm^{-3} and a viscosity number, determined according to ISO 307, of $VN = 64$ mL g^{-1}. The applied material exhibits a melting peak temperature of 179 °C and a crystallization peak temperature of 152 °C. To allow for thermographic investigations of exposure-induced temperature fields, a thermographic camera, VELOX 1310k SM, IRCAM GmbH, Erlangen, Germany, was used. Thermographic imaging was conducted using a frame rate of 355 Hz and a spatial resolution of 140 µm, covering a rectangular powder bed cross section of 40×40 mm^2. An emission coefficient of $\varepsilon = 0.9$ was specified for determining resulting temperature fields.

4.2. Design of Experiments

The applied methodology is based on complementary thermographic and morphological investigations of components manufactured at non-isothermal processing conditions.

Thermographic investigations are applied for characterizing thermal process properties of both conventional and quasi-simultaneous exposure strategies. For investigating processing properties of linear and fractal exposure strategies, single layers of square cross-sections exhibiting an edge length of 16.2 mm are fabricated. For enabling the comparison of linear and fractal exposure strategies at non-isothermal processing conditions, linear meander scanning and fractal scanning, corresponding to the Peano curve, are applied using a constant scan speed of 500 mm s^{-1}, a laser power of 16 W, a hatch distance of 0.2 mm, and a powder bed temperature of 100 °C. A single exposure step is applied to allow for the comparative assessment of processing properties of conventional exposure strategies under non-isothermal processing conditions. The edge length specified for the thermographic investigation of single-scan exposure strategies is chosen according to the underlying hatch distance of 0.2 mm, corresponding to the 4th iteration of the Peano curve.

In addition to the application of single-step exposure strategies, thermographic investigations are conducted for characterizing thermal process properties of quasi-simultaneous, fractal exposure by varying the number of consecutive scans for determining the influence of quasi-simultaneous processing conditions on resulting transient thermal fields. Manufactured specimens compromise cubic geometries with an edge length of 10 mm, allowing for the investigation of the thermal superposition of distinct segments by means of thermographic analysis. Specimens are manufactured using a hatch distance of 0.2 mm and a laser power of 2 W. For characterizing the influence of varying quasi-simultaneous conditions and varying powder bed temperatures on emerging temperature fields, a full-factorial parameter variation is conducted, as displayed in Table 1. Varied process parameters include the number of scans, corresponding to the number of consecutive exposure steps of a distinct segment.

Table 1. Variation of processing parameters applied for thermographic investigations of non-isothermal PBF.

Number of Scans	Powder Bed Temperature
10, 15, 20, 25	75 °C, 100 °C

Complementary morphological investigations of manufactured specimens are based on applying constant quasi-simultaneous parameters and varied thermal boundary conditions, corresponding to a powder bed temperature of 75 °C and 100 °C, respectively. Based on preceding thermographic investigations, a number of 25 consecutive exposure steps are applied to ensure a homogenous layer formation. To allow for the comparison of thermal and morphological material characteristics in dependence of the underlying process, geometrically identical specimens are manufactured by applying quasi-isothermal PBF using a powder bed temperature of 170 °C, a hatch distance of 0.2 mm, a laser power of 16 W, and an exposure speed of 2000 mm s^{-1}. Meander scanning is employed as the underlying exposure strategy.

4.3. Fabrication of a Complex Geometry Demonstrator

To investigate the capability of the proposed process strategy for the manufacturing of complex parts, a compression spring, exhibiting an outer diameter of 10 mm, a free length of 15 mm, and a wire diameter of 1 mm, is manufactured applying the Peano-based, quasi-simultaneous exposure strategy. In order to improve the manufacturing of thin-walled components with an increased surface-volume ratio, a quasi-simultaneous contour scan is applied subsequent to the hatch exposure. Applied processing parameters are displayed in Table 2.

Table 2. Exposure parameters of the fractal, quasi-simultaneous exposure of thin-walled components.

Number of Consecutive Contour Scans	Number of Consecutive Hatch Scans	Powder Bed Temperature	Laser Power	Exposure Speed
50	25	75 °C	2 W	500 mm s^{-1}

A contour offset of 0.2 mm is specified, corresponding to the applied hatch distance.

4.4. Thermal and Morphological Characterization

Thermal characterizations of manufactured specimens are conducted using differential scanning calorimetry (DSC), applying a heating rate of 10 K min^{-1}. Sample preparation for DSC measurements includes the removal of the edge region in order to allow for the comparability of varying processing conditions. Morphological characteristics are determined by applying both polarization microscopy and computed tomography. Polarized micrographs are prepared using thin sections of 10 µm thickness. Corresponding microscopic investigations are conducted using a Zeiss Axio Imager 2, Carl Zeiss Microscopy Deutschland GmbH, Oberkochen, Germany. Computed tomography of manufactured specimens is conducted using a sub-µ-CT, Fraunhofer Institute for Integrated Circuits (IIS) e.V., applying an isotropic spatial resolution of 4.5 µm. Subsequent analytical investigations of the spatial distribution of pores rely on script-based analysis of three-dimensional, binarized density distributions.

5. Results and Discussion

5.1. Limitations of Single-Scan Exposure Strategies

Considering processing properties of linear, conventional meander exposure at non-isothermal processing conditions, thermographic investigations of manufactured single layers exhibit considerable distortion following the exposure step. Displayed in Figure 6a, single layers manufactured using linear exposure paths depict a contraction perpendicular to the formed thermal gradient. Concerning the emergence of warping and curling of manufactured single layers, the manufacturing of parts without the application of support structures is inherently restricted due to build process interruptions. In contrast, the application of the fractal Peano strategy displays a considerably reduced extent of distortion, indicating an improved applicability of fractal exposure strategies for support-free non-isothermal processing.

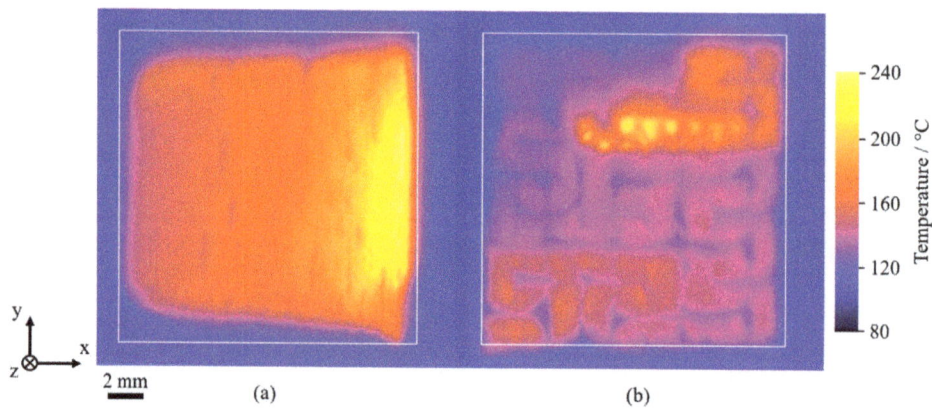

Figure 6. Thermographic imaging of the spatial distortion of the single-scan meander strategy (**a**) and the fractal Peano strategy (**b**) at a powder bed temperature of 100 °C, specified geometry indicated by a white square.

Considering reduced peak temperatures observed for the application of fractal exposure strategies [29], a reduced thermal superposition of consecutively exposed scan paths is correlated with the observed, reduced distortion of manufactured layers. A reduced resulting distortion is in accordance with findings described by Ma et al. [27] and Catchpole-Smith et al. [28] for PBF-LB/M, indicating a significant reduction of residual stresses [27,28] and the occurrence of stress cracking [28] of nickel-base superalloys. Consequently, fractal exposure of distinct segments allows for the considerable reduction of stress-induced distortion, hence representing a foundation for the proposed approach of combined fractal, quasi-simultaneous exposure.

5.2. Thermographic Analysis of Fractal, Quasi-Simultaneous Processing

The thermographic investigation of quasi-simultaneous exposure at non-isothermal conditions displays a characteristic gradual increase of observed mean segment temperatures. An increasing number of consecutive scans is correlated with a degressive increase, displayed in Figure 7a. Considering emergent peak temperatures, the excess of superficial melting of the powder bed significantly depends on the number of consecutive scans, being correlated with the applied energy density. Considering a degressive thermal increase, a similar thermal evolution can be observed for isothermal exposure using the meander exposure strategy due to a thermal superposition of consecutively exposed scan paths [19,20]. The influence of the applied powder bed temperature is clearly displayed with regard to the cooling rate, indicating a negative correlation of the applied temperature and the cooling rate. However, no significant influence of the powder bed temperature on resulting peak temperatures is observed, indicating a superposing influence of the local powder bed morphology on measured peak temperatures. Varying numbers of applied consecutive scans of distinct segments exhibit a quasi-linear relation of the number of applied scans and resulting peak temperatures, shown in Figure 7b, thus enabling the adjustment of resulting peak temperatures.

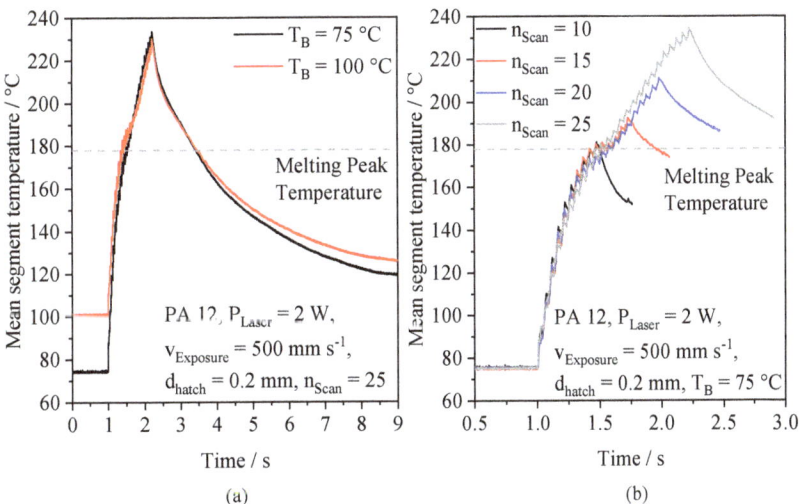

Figure 7. Influence of varying powder bed temperatures (**a**) and varying scan numbers (**b**) on resulting mean temperatures, occurring for the exposure of single, consecutively exposed segments.

Considering the influence of thermal interdependencies of consecutively exposed segments on resulting residual stress in PBF-LB/M [22] and the occurrence of thermal interaction of subsequently exposed scan vectors, observed in PBF-LB/P [20], correlations of fractal sequencing on resulting peak temperatures are investigated. Measured segment-specific peak temperatures of square 10×10 mm^2 cross-sections exhibit a merely insignificant influence of the exposure sequence, displayed in Figure 8.

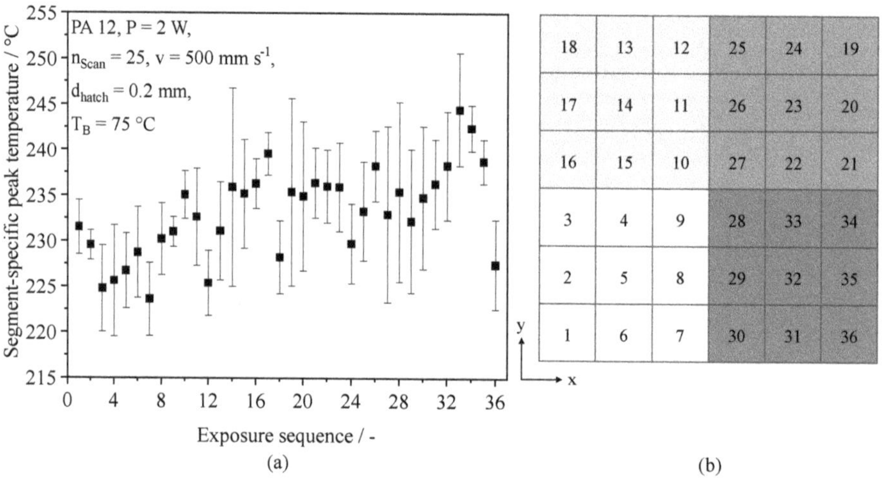

Figure 8. (a) Influence of the exposure sequence of a 10×10 mm^2 cross-section on resulting peak temperature values; (b) Spatial representation of the fractal exposure sequence of distinct segments.

Given the applied process parameters, fractal sequencing allows for the reduction of thermal superposition with regard to observed peak temperatures. Considering the fractal structure of the applied exposure sequence, a local, significant increase of peak temperatures can be observed, being correlated with the local meander-shaped sequence of the Peano curve. However, thermal superposition is predominantly limited to the local thermal interference of consecutively exposed segments, restricting the cross-component thermal superposition of exposed segments. Given the considerable reduction of part warping, a reduced thermal interdependence of distinct segments can be correlated with reduced stress-induced distortion, thus representing the foundation for the support-free, non-isothermal manufacturing of solid parts. Furthermore, with regard to the applied fractal sequencing of distinct segments, the observed merely local thermal superposition implies the possibility of limiting geometry-dependent influences on manufactured parts, hence allowing for the fabrication of parts with negligible variations of the local thermal history, regardless of underlying part geometries.

5.3. Part Morphology

Non-isothermal, quasi-simultaneous PBF of Polyamide 12 exhibits considerable morphological differences compared to components processed by means of isothermal PBF. Thin sections of manufactured cubic samples display a layer-wise crystallization of distinct layers, exhibiting a parabolic melt pool geometry of exposed segments, displayed in Figure 9.

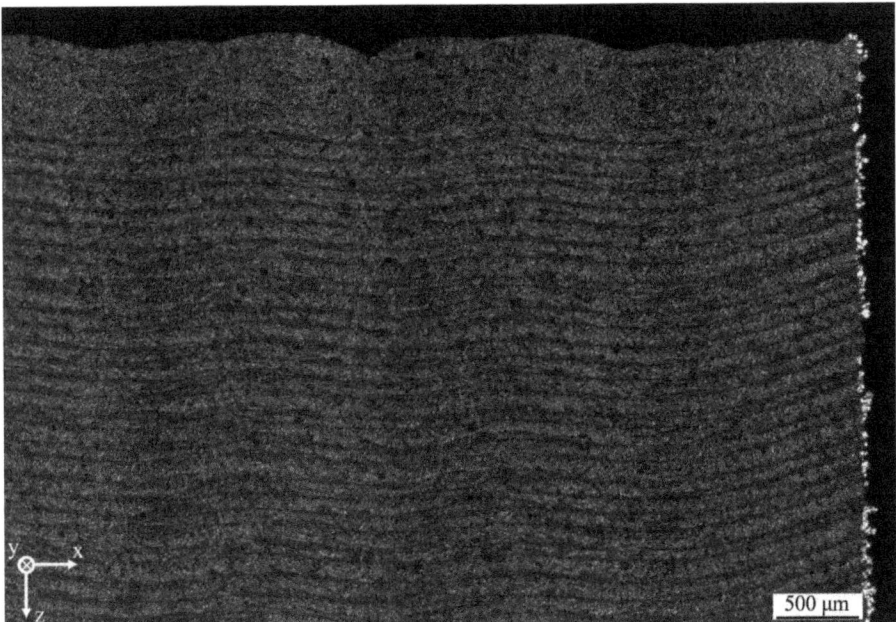

Figure 9. Polarization micrograph of a cubic specimen manufactured at non-isothermal conditions, $d_{\text{thin-section}} = 10$ µm.

Based on the displayed layer-wise structure, a dependence of the thermal penetration depth on the thermal superposition of distinct segments can be observed, leading to a reduced thermal penetration depth in the edge area. With regard to the observed, locally varying penetration depth, the previously observed relation of the number of consecutive exposure steps and the resulting peak temperature indicates a non-linear influence of the number of applied consecutive exposure steps. Considering the relation of the thermal diffusion length μ, the thermal impact time t, and the thermal diffusivity α [32], valid for the assumption of pulsed heating, a non-linear influence of a varying, exposure-dependent thermal impact time on the resulting thermal penetration depth is evident.

$$\mu = 2\sqrt{\alpha t} \tag{4}$$

In addition, the observed degressive increase of the segment-specific temperature indicates a superposing influence of the previously described relation of the number of scan paths and the resulting peak temperature on the emerging melt pool geometry. Consequently, the observed thickness of the top layer can be correlated with the applied quasi-simultaneous exposure, being induced by a considerable increased thermal impact time of the laser exposure compared to conventional exposure.

With regard to variations of the applied powder bed temperature implicitly influencing transient temperature fields of distinct segments, an influence on the extent of thermal degradation can be observed, displayed in Figure 10.

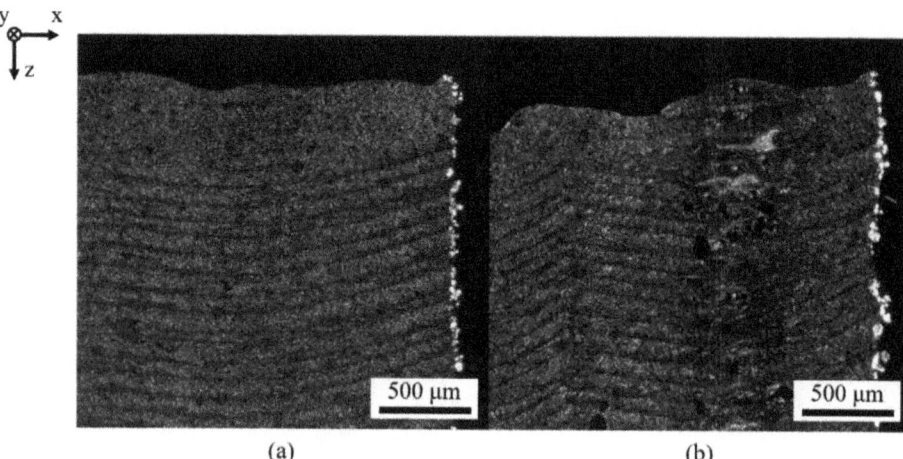

Figure 10. Polarization micrographs of cubic samples manufactured applying a powder bed temperature of 75 °C (**a**) and 100 °C (**b**).

Resulting morphologies depict a considerable influence of the applied powder bed temperature, affecting both the homogeneity of formed layers and the occurrence of thermal degradation in the center of distinct exposed segments. The extent of the occurrence of defects can be correlated with thermographic findings, showing a negative correlation of the applied powder bed temperature and the corresponding cooling rate, hence limiting thermal degradation. Furthermore, an increased powder bed temperature is correlated with the emergence of warping of manufactured parts, indicating the requirement of accelerated cooling of distinct exposed segments for increasing the geometric part accuracy.

Varying spherulite sizes, formed within distinct layers, imply the occurrence of locally varying thermal conditions within the melt pool, leading to varying crystallization kinetics and resulting spherulite sizes [33]. Considering crystalline structures formed in isothermal PBF-LB/P, non-isothermal PBF-LB/P promotes the formation of significantly reduced spherulite diameters regardless of the position within a particular layer. Therefore, specimens manufactured by means of non-isothermal PBF exhibit a reduced morphological homogeneity resulting from locally varying cooling rates, displayed in Figure 11, which is considered inherently related to non-isothermal processing conditions [34]. In contrast to isothermal processing, no structural morphologic differences of edge regions are formed in parts produced by means of non-isothermal processing, indicating a potential improvement of superficial mechanical and tribological properties of manufactured components.

Considering the influence of locally varying cooling rates on resulting spherulite diameters [33], similar spherulite dimensions observed within the build plane indicate a predominant thermal homogeneity during the exposure of distinct segments.

Figure 11. Polarization micrographs of Polyamide 12 specimens manufactured applying isothermal (T_B = 170 °C, v = 2000 mm s^{-1}, P = 16 W, d_{hatch} = 0.2 mm) (**a**) and non-isothermal (T_B = 75 °C, v = 500 mm s^{-1}, P = 2 W, d_{hatch} = 0.2 mm) (**b**) processing.

Using computed tomography, the occurrence of local thermal degradation is predominantly limited to the center of distinct segments. The formation of spherical pores can be correlated with excessive energy input into the melt pool, indicating a reduced level of optimum energy density. However, considering an overall density of 98.51% ±0.31, the satisfactory manufacturing of dense components can be demonstrated. Resulting spatial porosity distributions exhibit a periodic occurrence of thermal degradation within the build plane. In contrast, the average porosity level exhibits a predominant invariance towards the build height, displayed in Figure 12.

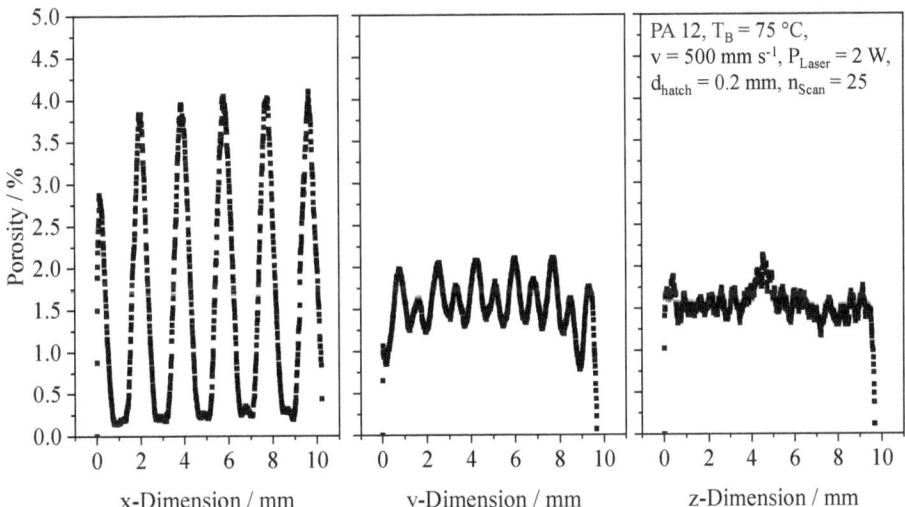

Figure 12. Spatial porosity distribution of a cubic sample, applying fractal sequencing of sub-segments, $d_{Smoothing}$ = 0.2 mm.

The applied exposure sequence of distinct segments is implicitly represented in resulting porosity distributions, displaying a considerable variance of porosity observed in the YZ-plane. Observed anisotropic influences are correlated with the exposure sequence of consecutive segments. With regard to the applied exposure sequence locally resembling the meander strategy, a thermal superposition of consecutively exposed segments can be correlated with evenly distributed porosity, induced by thermal decomposition. In contrast, an increased time span between the exposure of segments located next to each other limits thermal decomposition in the edge region of distinct segments, leading to the formation of dense part regions. These findings are in good accordance with thermographic results, indicating a local thermal superposition, thus influencing the local emergence of thermal decomposition. Resulting from the segmented structure of the exposure strategy, observed morphological properties indicate an implicit compensation of the shrinkage of distinct segments due to the segmented build process. Corresponding implications on a potentially improved dimensional accuracy of manufactured parts will be addressed in future research.

Therefore, in contrast to quasi-isothermal PBF, non-isothermal PBF allows for manufacturing quasi-dense sections, exhibiting an anisotropic, lamellar morphology. However, the locally restricted formation of porosity implies the requirement for parameter optimizations to avoid thermal degradation.

5.4. Thermal Material Properties

Using differential scanning calorimetry, similar properties of components manufactured using isothermal and non-isothermal PBF-LB/P can be observed. In contrast to the isothermal crystallization of Polyamide 12 at reduced ambient temperatures, described by Paolucci et al. [14], no exothermal peak, correlated with cold crystallization, can be observed for the application of a powder bed temperature of 75 °C, displayed in Figure 13. Displayed results are in good accordance with findings described by Zhao et al. [13], exhibiting a negative correlation of the applied cooling rate and the melting peak temperature of fabricated samples, determined by means of DSC.

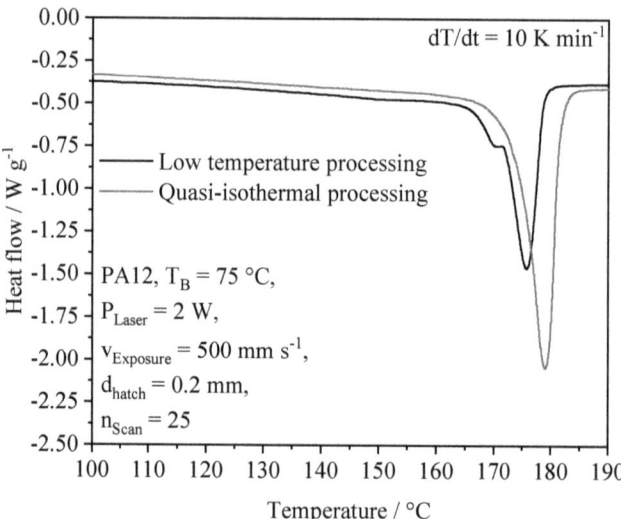

Figure 13. Differential scanning calorimetry of specimens manufactured using non-isothermal PBF and isothermal PBF.

The absence of cold crystallization indicates the predominant crystallization above the applied powder bed temperature, which is in accordance with thermographic investigations, showing a slow cooling process compared to thermal quenching, applied by Paolucci et al. [14]. Considering a melting enthalpy of -53.41 J/g $\pm\ 1.78$ and -61.95 J/g $\pm\ 1.35$ for Polyamide 12 processed at non-isothermal conditions and by means of quasi-isothermal processing, respectively, structural differences can be determined, indicating a slightly reduced crystallinity. Furthermore, a reduced melting point of specimens prepared at non-isothermal conditions is evident. These observations are in accordance with findings described by Kigure et al. [34] for the non-isothermal processing of Polyamide 12 using support structures, showing a comparable reduction in the degree of crystallinity and the melting peak temperature. Considering the applied powder bed temperature, no significant influence of an elevated powder bed temperature of 130 °C, described by Kigure et al. [34], can be observed. Therefore, the formation of similar thermal properties indicates a subordinate influence of the powder bed temperature on formed crystal modifications and the degree of crystallinity.

However, despite observed marginal variations in thermal material properties, non-isothermal processing promotes the formation of similar thermo-structural properties, correlated with similar crystalline structures, compared to the application of quasi-isothermal PBF-LB/P.

5.5. Manufacturing of Complex Geometries

The applicability of low temperature PBF of complex geometries is correlated with varying layer times [3], implying varying cooling times. Manufactured compression springs, displayed in Figure 14, demonstrate the suitability of the proposed fractal, non-isothermal processing strategy for successfully manufacturing geometries with varying cross-sections at non-isothermal processing conditions. Consequently, the proposed approach allows for the manufacturing of thick-walled and thin-walled components based on the combination of locally quasi-simultaneous exposure and fractal scan path generation. With regard to the geometric accuracy, structural restrictions can be observed.

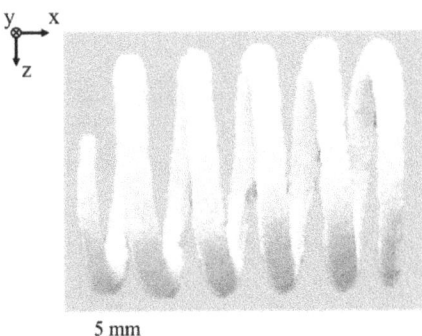

Figure 14. Depiction of a compression spring, manufactured by means of support-free low temperature PBF, $T_B = 75$ °C.

Resulting geometric deviations are correlated with the previously discussed elevated layer thickness, induced by the application of quasi-simultaneous exposure and an implicitly elevated thermal impact time. Consequently, optimizations of applied processing parameters and underlying thermal boundary conditions embed the potential for considerably increasing the geometric accuracy in future research.

6. Conclusions and Outlook

Within the present paper, a novel approach for the laser-based powder bed fusion of polymers was proposed. Based on the combination of fractal exposure strategies and quasi-simultaneous exposure of distinct segments, the non-isothermal, support-free manufacturing of varying part geometries could be demonstrated. Based on thermographic imaging, a gradual temperature increase of discrete segments of the exposed cross-section was identified. In combination with a subsequent immediate crystallization of distinct segments, the spatially and temporally discrete exposure constitutes the methodological foundation of non-isothermal processing. By applying considerably reduced powder bed temperatures, a lamellar morphology was obtained, exhibiting significantly reduced, locally varying spherulite sizes. Based on the segmented exposure strategy, shrinkage of discrete segments is limited to the mesoscopic level, allowing for the manufacturing of dense components.

Future research will focus on investigating mechanical properties of parts manufactured using low-temperature powder bed fusion, the optimization of process parameters in combination with the targeted local modification of the part morphology, as well as the integration of thermally sensitive additives, such as medically active substances, into manufactured parts. Furthermore, reducing the necessary number of consecutive exposure steps is essential for decreasing the layer time for enhancing the economic viability of low temperature PBF-LB/P. Therefore, potentials of low-temperature PBF-LB/P include the application of polymer blends with divergent thermal properties, considerably reduced thermal aging of polymers, and the manufacturing of thermally sensitive multi-material components.

Author Contributions: Conceptualization, S.S.; Software, S.S.; Investigation, S.S.; Methodology, S.S. and S.G.; Writing—Original draft, S.S. and S.G.; Funding acquisition, Writing—Review and editing and Project administration, D.D. All authors have read and agreed to the published version of the manuscript.

Funding: Funded by the Deutsche Forschungsgemeinschaft (DFG, German Research Foundation)—Project-ID 61375930—SFB 814—"Additive Manufacturing" TP B03. We acknowledge financial support by Deutsche Forschungsgemeinschaft and Friedrich-Alexander-Universität Erlangen-Nürnberg within the funding program "Open Access Publication Funding".

Institutional Review Board Statement: Not applicable.

Informed Consent Statement: Not applicable.

Data Availability Statement: The data presented in this study are available on request from the corresponding author. The data are not publicly available due to ongoing research in this field.

Conflicts of Interest: The authors declare no conflict of interest.

References

1. Wudy, K.; Drummer, D. Aging effects of polyamide 12 in selective laser sintering: Molecular weight distribution and thermal properties. *Addit. Manuf.* **2019**, *25*, 1–9. [CrossRef]
2. Jain, P.K.; Pandey, P.M.; Rao, P.V.M. Effect of delay time on part strength in selective laser sintering. *Int. J. Adv. Manuf. Technol.* **2008**, *43*, 117. [CrossRef]
3. Pavan, M.; Faes, M.; Strobbe, D.; Van Hooreweder, B.; Craeghs, T.; Moens, D.; Dewulf, W. On the influence of inter-layer time and energy density on selected critical-to-quality properties of PA12 parts produced via laser sintering. *Polym. Test.* **2017**, *61*, 386–395. [CrossRef]
4. Neugebauer, F.; Hikhin, V.; Ambrosy, J.; Witt, G. Isothermal and non-isothermal crystallization kinetics of polyamide 12 used in laser sintering. *J. Therm. Anal. Calorim.* **2016**, *124*, 925–933. [CrossRef]
5. Drummer, D.; Greiner, S.; Zhao, M.; Wudy, K. A novel approach for understanding laser sintering of polymers. *Addit. Manuf.* **2019**, *27*, 379–388. [CrossRef]
6. Soldner, D.; Greiner, S.; Burkhardt, C.; Drummer, D.; Steinmann, P.; Mergheim, J. Numerical and experimental investigation of the isothermal assumption in selective laser sintering of PA12. *Addit. Manuf.* **2020**, *37*, 101676. [CrossRef]
7. Soldner, D.; Steinmann, P.; Mergheim, J. Modeling crystallization kinetics for selective laser sintering of polyamide 12. *GAMM-Mitteilungen* **2021**, *44*, e202100011. [CrossRef]
8. Shen, F.; Zhu, W.; Zhou, K.; Ke, L.-L. Modeling the temperature, crystallization, and residual stress for selective laser sintering of polymeric powder. *Acta Mech.* **2021**, *232*, 3635–3653. [CrossRef]
9. Soe, S.P. Quantitative analysis on SLS part curling using EOS P700 machine. *J. Mater. Process. Technol.* **2012**, *212*, 2433–2442. [CrossRef]
10. Nakamura, K.; Watanabe, T.; Katayama, K.; Amano, T. Some aspects of nonisothermal crystallization of polymers. I. Relationship between crystallization temperature, crystallinity, and cooling conditions. *J. Appl. Polym. Sci.* **1972**, *16*, 1077–1091. [CrossRef]
11. Nakamura, K.; Katayama, K.; Amano, T. Some aspects of nonisothermal crystallization of polymers. II. Consideration of the isokinetic condition. *J. Appl. Polym. Sci.* **1973**, *17*, 1031–1041. [CrossRef]

12. Ziabicki, A. *Fundamentals of Fibre Formation: The Science of Fibre Spinning and Drawing*; Science of Fibre Spinning and Drawing; Wiley: Hoboken, NJ, USA, 1976; ISBN 9780471982203.
13. Zhao, M.; Wudy, K.; Drummer, D. Crystallization Kinetics of Polyamide 12 during Selective Laser Sintering. *Polymers* **2018**, *10*, 168. [CrossRef] [PubMed]
14. Paolucci, F.; Baeten, D.; Roozemond, P.C.; Goderis, B.; Peters, G.W.M. Quantification of isothermal crystallization of polyamide 12: Modelling of crystallization kinetics and phase composition. *Polymer* **2018**, *155*, 187–198. [CrossRef]
15. Niino, T.; Haraguchi, H.; Itagaki, Y.; Iguchi, S.; Hagiwara, M. Feasibility study on plastic laser sintering without powder bed preheating. In Proceedings of the 2011 International Solid Freeform Fabrication Symposium, Austin, TX, USA, 8–10 August 2011; University of Texas at Austin: Austin, TX, USA, 2011; pp. 17–29.
16. Niino, T.; Haraguchi, H.; Itagaki, Y.; Hara, K.; Morita, S. Microstructural observation and mechanical property evaluation of plastic parts obtained by preheat free laser sintering. In Proceedings of the 2012 International Solid Freeform Fabrication Symposium, Austin, TX, USA, 6–8 August 2012; University of Texas at Austin: Austin, TX, USA, 2012; pp. 617–628.
17. Laumer, T.; Stichel, T.; Amend, P.; Schmidt, M. Simultaneous laser beam melting of multimaterial polymer parts. *J. Laser Appl.* **2015**, *27*, S29204. [CrossRef]
18. Chatham, C.A.; Long, T.E.; Williams, C.B. Powder bed fusion of poly(phenylene sulfide) at bed temperatures significantly below melting. *Addit. Manuf.* **2019**, *28*, 506–516. [CrossRef]
19. Wegner, A.; Witt, G. Process monitoring in laser sintering using thermal imaging. In Proceedings of the 2011 International Solid Freeform Fabrication Symposium, Austin, TX, USA, 8–10 August 2011; University of Texas at Austin: Austin, TX, USA, 2011.
20. Greiner, S.; Wudy, K.; Wörz, A.; Drummer, D. Thermographic investigation of laser-induced temperature fields in selective laser beam melting of polymers. *Opt. Laser Technol.* **2019**, *109*, 569–576. [CrossRef]
21. Greiner, S.; Jaksch, A.; Cholewa, S.; Drummer, D. Development of material-adapted processing strategies for laser sintering of polyamide 12. *Adv. Ind. Eng. Polym. Res.* **2021**, *4*, 251–263. [CrossRef]
22. Zou, S.; Xiao, H.; Ye, F.; Li, Z.; Tang, W.; Zhu, F.; Chen, C.; Zhu, C. Numerical analysis of the effect of the scan strategy on the residual stress in the multi-laser selective laser melting. *Results Phys.* **2020**, *16*, 103005. [CrossRef]
23. Chen, C.; Yin, J.; Zhu, H.; Xiao, Z.; Zhang, L.; Zeng, X. Effect of overlap rate and pattern on residual stress in selective laser melting. *Int. J. Mach. Tools Manuf.* **2019**, *145*, 103433. [CrossRef]
24. Hagedorn-Hansen, D.; Bezuidenhout, M.B.; Dimitrov, D.M.; Oosthuizen, G.A. The effects of selective laser melting scan strategies on deviation of hybrid parts. *S. Afr. J. Ind. Eng.* **2017**, *28*, 200–212. [CrossRef]
25. Lu, Y.; Wu, S.; Gan, Y.; Huang, T.; Yang, C.; Junjie, L.; Lin, J. Study on the microstructure, mechanical property and residual stress of SLM Inconel-718 alloy manufactured by differing island scanning strategy. *Opt. Laser Technol.* **2015**, *75*, 197–206. [CrossRef]
26. Yang, J.; Bin, H.; Zhang, X.; Liu, Z. Fractal scanning path generation and control system for selective laser sintering (SLS). *Int. J. Mach. Tools Manuf.* **2003**, *43*, 293–300. [CrossRef]
27. Ma, L.; Bin, H. Temperature and stress analysis and simulation in fractal scanning-based laser sintering. *Int. J. Adv. Manuf. Technol.* **2007**, *34*, 898–903. [CrossRef]
28. Catchpole-Smith, S.; Aboulkhair, N.; Parry, L.; Tuck, C.; Ashcroft, I.A.; Clare, A. Fractal scan strategies for selective laser melting of 'unweldable' nickel superalloys. *Addit. Manuf.* **2017**, *15*, 113–122. [CrossRef]
29. Greiner, S.; Schlicht, S.; Drummer, D. Temperature field homogenization by fractal exposure strategies in laser sintering of polymers. In Proceedings of the 17th Rapid.Tech 3D Conference, Erfurt, Germany, 22–23 June 2021; Carl Hanser Verlag GmbH & Co. KG: Munich, Germany, 2021; pp. 188–189, ISBN 978-3-446-47171-9.
30. Breinholt, G.; Schierz, C. Algorithm 781: Generating Hilbert's Space-Filling Curve by Recursion. *ACM Trans. Math. Softw.* **1998**, *24*, 184–189. [CrossRef]
31. Peano, G. Sur une courbe, qui remplit toute une aire plane. *Math. Ann.* **1890**, *36*, 157–160. [CrossRef]
32. Marín, E. Characteristic dimensions for heat transfer. *Lat.-Am. J. Phys. Educ.* **2010**, *4*, 10.
33. Piccarolo, S.; Saiu, M.; Brucato, V.; Titomanlio, G. Crystallization of polymer melts under fast cooling. II. High-purity iPP. *J. Appl. Polym. Sci.* **1992**, *46*, 625–634. [CrossRef]
34. Kigure, T.; Yamauchi, Y.; Niino, T. Relationship between powder bed temperature and microstructure of laser sintered PA12 parts. In Proceedings of the 2019 International Solid Freeform Fabrication Symposium, Austin, TX, USA, 12–14 August 2019; University of Texas at Austin: Austin, TX, USA, 2019.

MDPI
St. Alban-Anlage 66
4052 Basel
Switzerland
Tel. +41 61 683 77 34
Fax +41 61 302 89 18
www.mdpi.com

Polymers Editorial Office
E-mail: polymers@mdpi.com
www.mdpi.com/journal/polymers